# MANAGING RURAL DEVELOPMENT

INDO-DUTCH STUDIES ON DEVELOPMENT ALTERNATIVES–7

## Indo-Dutch Studies on Development Alternatives

The present study is the seventh volume in a series of studies sponsored under the *Indo-Dutch Programme on Alternatives in Development* (IDPAD). This programme grew out of intensifying contacts and cooperation in the seventies between Indian and Dutch social scientists willing to share each other's experiences in development research. Led by a common concern about the need to critically assess existing structures and policies from the point of view of meeting basic needs of large masses of people and to conceive alternatives that heighten the capabilities of the latter to serve as creative agents in the evolving world economy and society, a workplan of joint research was agreed on in 1980 forming the first phase of IDPAD (1981–1983).

Since its inception the programme is being jointly implemented by the Indian Council of Social Science Research (ICSSR), New Delhi, and the Institute of Social Science Research in Developing Countries (IMWOO), The Hague, and financed to a major extent by a grant from the Dutch government and a generous contribution from the Indian government. Following the second phase (1984–1989), which is based on the same principles, a third phase has been prepared (1990–1994) by the Joint Committee of IDPAD under the co-chairmanship of the late Professor S. Chakravarty, Delhi School of Economics, and Professor J. Breman, University of Amsterdam.

This series of *Indo-Dutch Studies on Development Alternatives* includes only those research results obtained under IDPAD which are considered essentially qualified for wider dissemination. However, the views expressed and facts stated therein are, of course, those of the authors.

1. H.H. de Haan: *Alternatives in Industrial Development: Sugarcane Processing in India* (1988).
2. Jairus Banaji and Rohini Hensman: *Beyond Multinationalism: Management Policies and Bargaining Relationships in International Companies* (1990).
3. M. Doornbos *et al.*: *Dairy Aid and Development: India's Operation Flood* (1990).
4. M. Doornbos and K.N. Nair (eds.): *Resources, Institutions and Strategies: Operation Flood and Indian Dairying* (1990).
5. Nirmala Banerjee: *Indian Women in a Changing Industrial Scenario* (1991).
6. Nirmal Sengupta: *Managing Common Property: Irrigation in India and the Philippines* (1991).

# MANAGING RURAL DEVELOPMENT

## Health and Energy Programmes in India

Edited by

**Hein Streefkerk and T.K. Moulik**

INDO-DUTCH STUDIES ON DEVELOPMENT ALTERNATIVES–7

**Sage Publications**
New Delhi/Newbury Park/London

*First Published in 1991 by*

**Sage Publications India Pvt Ltd**
M 32 Greater Kailash Market I
New Delhi 110 048

**Sage Publications Inc**
2455 Teller Road
Newbury Park, California 91320

**Sage Publications Ltd**
6 Bonhill Street
London EC 2A 4PU

Published by Tejeshwar Singh for Sage Publications India Pvt Ltd, phototypeset by Mudra Typesetters, Pondicherry and printed at Chaman Offset Printers, Delhi.

**ISBN** 0–8039–9678–0 (U.S.)
81–7036–220–2 (India)

# CONTENTS

# LIST OF ILLUSTRATIONS AND TABLES

## A. ILLUSTRATIONS



## B. TABLES

# PREFACE

The Indo-Dutch Programme for Alternatives in Development (IDPAD) is a collaborative effort of the Indian Council of Social Science Research (ICSSR), New Delhi and the Institute for Social Science Research (IMWOO), The Hague. The initiative for the Programme was taken by Dr. Jan P. Pronk, former and present Dutch Minister for Development Co-operation, in 1977 and it is funded principally by the Dutch Government and partly by the Government of India. Considerable preparatory work, especially two workshops in late 1978 and 1979 helped shape the Programme initiated in mid-1980. The basic aim of IDPAD has been increasing cooperation between Indian and Dutch scholars to analyse development processes in India, The Netherlands, other EEC and developing countries and to keep open the possibility of undertaking research, at a later stage, on more global, internationally oriented processes. IDPAD Phase-I consisted of a few selected research projects in the three chosen fields of health care, women studies and industrialisation. Phase-II, commenced in 1983, consists of a number of research projects in the fields of comparative Asian rural transformation, changing trends in European society and the New International Economic Order (NIEO).

Professor Enno Hommes, Co-Chairman of the First Joint Indo-Dutch Committee on Research Collaboration in Social Sciences, had taken the initiative in launching a Rural Technology Project under the Programme with the cooperation of Dutch colleagues and in collaboration with a team of experienced researchers at the Indian Institute of Management, Ahmedabad, under the leadership of Professor Tushar Moulik. The relatively high rate of success of this project owes largely to the skilful coordination by Professor Hommes and the fruitful collaboration between Dutch and Indian scholars.

The seminar on 'Management of Rural Development: Health and Energy Programmes' held under the auspices of the Institute of Development Studies (IDS) at Jaipur in March 1986, was the culmination of the continuing dialogue and debate among concerned Dutch and Indian scholars, with scholars from the IDS also contributing to the deliberations and debate. The editors of the volume, Hein Streefkerk and Tushar Moulik, and Klaas W. Van der Veen and Enno Hommes who have contributed the 'Introduction' and the 'Conclusions' have done a painstaking job in preparing the volume for publication. According to Van der Veen, two approaches to development programmes, namely, the management approach and the anthropological-sociological approach crystallised. However, Hommes would like to add three more approaches, viz., 'the technological', 'the institutional' and 'the span of control'. Whether some of these approaches could be subsumed under the earlier two is not examined.

I had the privilege of being associated with IDPAD for a number of years since its very inception as a senior functionary of the ICSSR and we were happy to host the seminar at the IDS and contribute to its deliberations. The volume presents a rich intellectual menu for further informed debate on a number of unresolved issues.

Last but not least, thanks should go to Ms. Ada Tieman and Ms. Maartje Uneputty of the Department of South and Southeast Asian Studies of the University of Amsterdam, who have been of great help in preparing the manuscript.

*Honorary Senior Fellow,* D.D. NARULA
*Institute of Development Studies,*
*Jaipur.*

# 1

# INTRODUCTION: MANAGEMENT AND THE SOCIO-CULTURAL CONTEXT

**Klaas W. van der Veen**

Most of the chapters in this book are the outcome of short exploratory investigations into the implementation of new technologies in rural areas in India. They concern technologies in the fields of public health and alternative energy. Private development organisations and the Indian government are responsible for their implementation.

Most of the chapters are based on studies conducted by Indian and Dutch scholars in Gujarat. The Indian-authored chapters, except for one, are largely based on secondary sources. The Dutch-authored chapters are based on information collected during fieldwork. The studies were carried out within the framework of the Indo-Dutch Programme for Alternatives in Development (IDPAD). The researchers had a common though broad frame of reference, focusing on the relationship between the type of project management and the results of the programmes.

Contacts between researchers at the Indian Institute of Management in Ahmedabad and their counterparts at the Twente University of Technology, the Royal Tropical Institute in Amsterdam and the University of Amsterdam, were at the root of the common point of departure. Initially, the Twente University and the Indian Institute of Management collaborated in a study on the implementation of rural technology in Gujarat and Tamil Nadu (see Blankenberg's paper in this volume). After a few years, other Dutch scholars joined this coordinated research effort.

The Indian and Dutch researchers had both been disappointed by the outcomes of various development programmes in their own work in South Asia which was also experienced by others (Morss and Hatch *et al.* [1975]; Paul [1982]; Rondinelli [1983]). They agreed with the diagnosis that socio-economic conditions are the main causes of these failures. However, some of them argued that the type of management of the programmes, or rather the lack of it, should also be considered as an important determinant for success or failure of development projects. This view led to the formulation of a research project consisting of separate investigations into different development programmes and aspects of their management.

The Dutch colleagues who took part in the initial discussions were skilled researchers and regular visitors to India and Gujarat.[1] They were able to contribute to the project for only short periods of five to six weeks. The time restrictions entailed a pragmatic selection of research subjects and an adjusted methodology. To obtain the highest returns from short research visits the studies had to be tied directly to the experience of the researchers. Accumulated knowledge and acquaintance with the 'region' determined subjects and location of the studies. Thus, rural health, rural drinking water supply, afforestation and renewable energy were chosen as the research topics, and Gujarat was the main research area. This selection procedure guaranteed optimum use of knowledge and the commitment of the researchers as they could freely choose their subjects and the methods to study them. It meant, however, that a common methodology for the different investigations of the project became difficult to realise. A comparison between programmes implemented in 'central' and 'peripheral' regions of Gujarat was one of the original methodological principles. However, due to time constraints only two researchers were able to do this.

Furthermore, intensive analysis of the effects of management procedures necessitates equal attention being given to all administrative strata such as the international, national, state, region or village levels, where important decisions are taken or executed. Unfortunately, this was also not possible for the same reason.

As the studies progressed it became increasingly apparent that, despite a common frame of reference, two different perspectives were crystallising: the management approach to development programmes; and the anthropological-sociological study of development projects and their results. Different opinions about the ability to control influences from the 'environment' on the implementation of development programmes are a major point of disagreement between the two perspectives.

The advocates of the management approach consider strategic management an essential part of the management of programmes as it determines, to a large extent, the performance of development projects. Strategic management refers to decisions about the development of goals, objectives and major policies for the programme and their adaptation to changes either in the external environment, or the internal organisation, or both. Processes of strategic management differ from other management processes by their potential of influencing the future course of programmes. Strategic decisions are generally non-repetitive in character and are the responsibility of management at the highest level.

The 'environment', the 'organisation' and the long-term 'goals and objectives' of a programme are the three variables which influence organisational performances according to the strategic management approach. The term 'environment' is used for the conglomeration of technological, regulatory, social, economic and competitive forces external to a development programme. 'Organisation' refers to the durable arrangements within a programme to perform the given tasks, and to the systems of information, planning and monitoring. The latter includes coordination, training and evaluation of the programme and personnel, in particular leadership and its influence on the process of management. The goals and objectives of a programme, or its 'strategy', are a third major determinant of the programme's performance.

Decisions of strategic management combine the three variables in order to achieve the objectives of the programme. The argument is that when choices of 'organisation' and 'strategy' are consistent with the characteristics of the 'external environment', they are likely to lead to superior performances. The choices pertain to the ideology and goals underlying a project; the organisations executing the project; needs assessments; and the monitoring and evaluation of programmes (Chaudhury [1985]; Moulik and Chaudhury [1986]).

The concept of strategic management has been developed through a large number of case studies of private enterprises originally in the USA and later in various European countries. It is a normative concept and suggests universal applicability such as that claimed, for instance, for orthodox mono-economic theory. According to this theory, 'economics consists of a number of simple yet "powerful" theorems of universal validity' (Hirschman [1981]). Similarly, if the strategic management of a programme is handled properly, success is expected, irrespective of the specific social environment and its influences, which can be controlled.

The advocates of the sociological-anthropological view of development projects do not deny the importance of proper management procedures. But they perceive the relation between the project and the 'environment' differently. In their opinion, projects and their results reflect the socio-economic conditions and express the dominant structural and cultural characteristics of the society in which they take place. The structural variables comprise social divisions like caste and class, the contradictions between rich and poor, and between the 'centre' and the 'periphery'.

Cultural characteristics include ideas and codes about interpersonal relations, hierarchy, the distinction between men and women, and so on.

The chapters in this volume make a differential use of these two perspectives differing as to the scope of generalisations and importance given to local conditions. The articles by Blankenberg and Satia emphasise the strategic management approach. Skutsch and Streefland combine a management perspective with local level data. Ahuja *et al.*, Schenk-Sandbergen, Streefkerk, Van der Veen and Frank de Zwart attach great weight to the socio-economic environment.

The controversy between the advocates of the two perspectives is not resolved in this volume but will continue to provide a framework for extensive further discussion. Professor Iqbal Narain (former Member Secretary of the Indian Council of Social Science Research) raised the issue during the Indo-Dutch Jaipur Conference on the management of rural technologies in 1986 when the results of the project were presented. He expressed doubts about the general applicability of the management concept for development programmes in rural India. He pointed to the two-way interaction between management and rural social realities.

Iqbal Narain argued that Gujarat, which provides the bulk of the research material, is a special case in India 'because an average Gujarati has a managerial cultural orientation which can be easily differentiated from managerial orientations elsewhere in India'. The managerial orientation of Gujaratis, again, differs from the average Dutch management style. The multi-cultural composition of the research group therefore was advantageous as it provided a means of comparison.

Regardless of the various positions on the capacity to control influences from the 'environment', the socio-economic context is an important variable which figures significantly in almost all chapters. Hence, structural and cultural aspects with common relevance for the majority of the chapters will be discussed in the following section.

First, Gujarat's dominant structural features are highlighted. Next, the cultural codes, i.e., the shared beliefs and common meanings are presented. Finally, the case studies are discussed in relation to these structural features and cultural codes.

**Main features of Gujarat's social structure**

In 1985, the state of Gujarat had an estimated population of 38.5 million people, of which 69 per cent lived in villages (Statistical Outline [1987]). Gujarati society is characterised by sharp inequalities: between the very prosperous (central) areas and the very poor (peripheral) regions and in each of these, between rich and poor people. Gujarat is among those Indian states where capitalist modernisation of agricultural and industrial production is becoming dominant. This process is accompanied by increased social differentiation and marginalisation of large segments of the rural population.

Despite these tendencies, the emerging social formation cannot be described and analysed simply as a class society. Distinctions are expressed in the idiom of caste and coloured by hierarchical notions of status and the necessity of conforming to rules of proper behaviour as expressed in 'sanskritisation'.[2] Thus, social reality can be perceived in terms of hierarchy, social division, and group identification. The local society is structured along a hierarchical division of castes or jatis, the endogamous groups based on caste distinctions. Endogamy serves to preserve hierarchy and division, and to maintain identification with the group whose members share a common lifestyle. In the villages these groups occupy a clearly demarcated and easily discernible space. The dominant castes reside in the centre of the villages, the low and poorest jatis or tribal groups occupy separate districts or wards on the outer fringes of the villages. Similar strong features of group identification hold for 'tribals' and members of different religious communities.

Other relevant structural features are the characteristics of public administration and the relation between public and private spheres. Many of the development projects are official programmes initiated and managed by the government, which can be described as 'a grace and favour state' (Wade[1985]). Patronage and the politics of factionalism are among the main structural features of such a polity. Public servants see themselves as dispensers of favours to which clients try to make themselves entitled. This aspect is of particular significance because of the prominent role of the government in the Indian economy. It is a mixed economy with almost all agricultural production realised by peasants and farmers. A large part of industrial and commercial assets is owned by private entrepreneurs. The state also controls many economic resources and is an industrial investor. It tries to regulate socio-economic life through bureaucratic rules and regulations, or rather through discretionary controls (Myrdal [1970]). Furthermore, there is a proliferation of ministries, departments and other organisations involved in the organisation, planning, and execution of rural development activities. This results in a lack of focused, well structured, and coordinated efforts thereby providing ample opportunity for officers to interfere widely in public and private programmes.

The division of power between central government and local authorities is another important administrative feature. Local self-government, called panchayati raj,[3] and the concentration of power in the hands of a few king-pin bureaucrats, such as collectors in the districts and joint secretaries in the ministries, create a complex game of balancing different forces and an intricate division of spoils. Decentralisation both in the administration and in the political structure has contributed to the complexity of the system of horizontal and vertical loyalties. Local politicians do have relationships with the administrators at the taluka level, but at the same time they are linked to the loyalty network of the Members of the Legislative Assembly (MLA) in Ahmedabad, the state capital. Some programmes are implemented through the taluka administration, others are the responsibility of the officers at the district headquarters or of authorities in Ahmedabad or even Delhi. Administrative and political decentralisation increases local participation, but it also gives local elites great scope for influencing the content and implementation of development programmes.

Finally, the increasing number and role of non-governmental organisations (NGOs) must be mentioned. According to Iyengar ([1987], p. 83), there were 805 voluntary organisations in Gujarat in 1986. Social work has a long tradition among the Gujarati elite. Gujarat was Gandhi's native state and many social workers and their organisations in the rural areas still work according to his principles (Shah and Das [1988], p. 88). They played, and still play, a key role in local social and political life and have a stake in the local status quo. However, when they commit themselves to eliminate social abuses they are bound to collide with vested interests.

### Characteristics of Gujarat's cultural code

The important role of cultural factors in development projects necessitates a discussion of the main features of the local cultural code. A detailed analysis is even more necessary as the assumptions of the management model originated in a cultural context different from that of India.

This review must start with an important methodological remark. Salient cultural features will be singled out and opposites emphasised. They are 'ideal types', mental constructs meant to clarify a complex reality.

Culture, or a cultural code, is a system of shared meanings, concepts and rules that underlie, and are expressed in everyday behaviour. A cultural code is not fixed and unalterable, nor is it the same for all members of a society. Its prescriptions differ for men and women, young and old, rich and poor, Brahmin and Harijan.

Central to the present argument is the basic ambiguity of people as individuals and people as groups. Elias [1978, p. 228] points out that 'concepts such as "individual" and "society" do not relate to two objects existing separately but to different yet inseparable aspects of the same human beings.' This ambiguity creates the opportunity to emphasise one or the other dimension, i.e., 'individuality' or 'group orientation'. The complexity and articulation of the different perspectives is obviously related to the social structure of societies.

A group oriented cultural code accentuates the notion that people are bound to others whom they cannot choose. The individualist perspective, on the contrary, emphasises the ideal that people can choose partners for different types of bonds.

These different perspectives imply different types of rationality. The dialectics of ambiguity force people to express their attachments to others differently and to use other methods to serve their own ends in a social context with a cultural code that emphasises inescapable interpersonal attachments, compared with a context and code stressing individual freedom of choice. In this volume the main interest lies in the confluence of these two ideal types, as will be clarified later.

In South Asia, group orientation is emphasised, while in the West the ideal of individual choice predominates. The ascendancy of the individualist view in the West was part of the rise of capitalism. It implied the growth of a specific type of rationality. Weber was among the first to point systematically to these tendencies. He distinguished between traditional or 'Wert-Rationalität' (value rationality) and a modern or capitalist 'Zweck-Rationalität' (means-end rationality). In terms of the present analysis, value rationality can be seen as emphasising the importance of ascribed interpersonal bonds as a major yardstick in evaluating the rationality of decisions. Weber also pointed to the growth of impersonal administration in modern bureaucracies as part of the rise of capitalism and the emergence of its specific rationality. The study of public and private organisations and the generation of management models since the thirties is a manifestation of the same development.

An emphasis on individual freedom, a precise definition of the conditions of interaction, contractual relations and a sharp demarcation of the 'private' and 'public' spheres of life, are the main social implications of a cultural code that emphasises individuality.

South Asian cultural perspectives stress the subordination of the individual's interest to those of the group; dependability is accorded high priority. The structural diversification of society into numerous endogamous groups such as jati (caste) or tribal groups or wider frames of primordial loyalties such as native area, regional origin, or religious denomination constitute the structural counterpart of the group oriented code.

The particular way of pursuing individual goals and the ensuing rationality implied by this code can best be described as 'invest(ment) in people', to use the apt characterisation of Srinivas [1955]. The maxim 'to oblige and to be obliged' is the prevalent social process. A diffusion of claims and counter-claims, few of which are settled immediately, provides optimal conditions for achieving individual aims within the limits of a dominant group orientation.

The encompassing nature of these types of relations makes it necessary to control them. Hierarchy, status articulation, and redistribution provide the means of doing so. These structural features belong to a single family of phenomena best described as patronage. Through patronage, mutual obligations between people of unequal status can be effectuated and scarce resources redistributed; it is at the same time a means of control. The following example illustrates the principle. A patron, say the president of the village panchayat, invests in people by promising or distributing favours to gain or legitimise his political power. The clients, in turn, show their loyalty and try to capitalise on it. However, the patron is able to control claims and counter-claims by initiating or denying them, because of his power. Hierarchy governs redistribution in another way too. The patron's loyalty to his equals, his caste fellows, who generally are his class fellows too, overrules his duties toward social and economic inferiors. Finally, there are far more clients than patrons, which often enables patrons to ignore the individual claims of their dependents.

The growing alienation of large sections of the rural population from circuits of redistribution is a dominant structural process in Gujarat and other rural areas of India. It is the social cause of large-scale rural marginalisation, which is among the main structural features discussed earlier.

### Cultural codes and strategic management

In this final section the impact of cultural codes on the management of technology programmes will be discussed. This discussion is necessitated by the Western, or rather American, orientation of organisational models applied to Indian society since the introduction of community development programmes in 1954. It must be stressed that different codes produce different kinds of responsibilities and commitments. The individual centred perspective emphasising personal freedom necessitates rules specifying claims and counter-claims, services and payments, demands and duties, and detailed instructions as to task performance and legal procedures. Commitment to tasks and adherence to rules are conditions for personal satisfaction. The task group provides people with a sense of belonging and with chances for occupational success. Commitment to tasks implies shared responsibility for a common cause, though this responsibility is graded according to the administrative hierarchy.

The group code stresses interpersonal bonds, diffusion of claims and counter-claims, and achieving one's aims through the intercession of others, preferably powerful patrons. In the Indian structural context of inequality and its ideological justification, dependability is the undisputed liability of those whose lifestyle and position enable them to bear it. Commitment means commitment to people to whom allegiance is owed and whose dependability, the willingness to accept responsibility, is taken for granted. Involvement with, and responsibility for, a common aim depends on the shared expression of loyalties. However, the vast socio-economic differences in South Asia often make these ideals more nominal than real.

In present-day South Asia, group oriented perspectives and individual centred codes have merged since the enormous expansion of government administration in the fifties. In this period, particularly after the introduction of the community development programme in 1954, the conflict between modern development oriented administration and traditional bureaucracy became evident. The main functions of the traditional bureaucratic system were collecting tax and maintaining law and order. The system was moulded by the local value system during the long colonial period. Community development administration was a modern management structure with a strong American background. It was based on the idea that 'individuals fight for their rights' through voluntary cooperative actions, which is not in line with local customs. The integration of the two democratic cultures with their conflicting value systems is, to a certain extent, reflected in the problems of integrating, or rather accommodating, them in the minds and behaviour of individual administrators.

Bureaucratic controls and impersonal procedures operate in conjunction with group oriented ideals. The confluence of two cultural perspectives generates responsibilities and commitments that interfere with the successful implementation of rural development programmes. A main contradiction exists between the beneficiaries' perception of patronage and the administrative guidelines of officers. Officers must meet targets to safeguard their position in the government bureaucracy. Beneficiaries on their part, perceive official programmes, like the distribution of stoves or piped drinking water, as favours. They consider their own muted acceptance of these favours as an act through which they oblige officials. This is why broken stoves, leaking water pipes, the failure to immunise children, and other discomforts are not reported, and often are the object of neither criticism, nor protest. Patronage means submitting to patrons, not irritating them. The public's commitment is to the patrons and not to the programme. The involvement of beneficiaries in various programmes is also hampered by the emotional and physical distance between them and the administrative personnel. High officials in particular keep relations impersonal and allow little opportunity 'to oblige and be obliged', unless by outright bribery or corruption. This generally implies relations between officers and the village elite.

At the other end, on the officials' side, commitment is also a problem. When villagers fail to react, officials can easily restrict programmes to the mere provision of facilities and to reaching targets, without involving beneficiaries on a long-term basis.

Problems of commitment and responsibility seem to be less acute when NGOs are involved in the implementation of programmes. Their workers' cultural perspectives are often more attuned to the particularistic expectations of the villagers. Furthermore, programmes are the main sources of livelihood for their workers. Therefore the continuation of a project is their major concern and enhances their commitment and responsibility.

The distinction between public and private development organisations is important. The 'favours' that government officers bestow upon the general public are considered as second rate, meant for the poor or people who do not have their 'own' good patrons. However, this does not always mean that 'private is good and public is bad'. This is illustrated by the example of the TB programme, analysed by Van der Veen in this volume.

Private practitioners and government organisations provide therapy to cure tuberculosis. Prolonged chemotherapeutical treatment for a period of one year is the only guarantee of complete recovery. The services of private doctors must be paid for, while government treatment is free. Ironically, because patients think they are treated better by private doctors, the private sector attracts many more patients than the state services. Patients thus minimise their chances of complete cure because the majority of them cannot afford to pay for the full length of treatment. The unfortunate attraction of private doctors is based on the combination of different cultural perspectives and rationality, and on the initial quick alleviation of TB symptoms after the introduction of chemotherapy. Patients expect proper attention and successful treatment, for they have obliged the doctor, their patron, by paying him. For the doctors, profit is what counts. They do not feel obliged to poor patients and never tell them that a long and expensive treatment is needed. The doctors are afraid of losing patients and are not dependable.

Public and private organisations do not always compete economically. But they certainly do compete on another plane: both are eager to create adherents and support. Or as Skutsch writes: 'Any programme is its own life and not the real output . . . the true purpose of a programme is to survive and to strengthen the position of various parties involved in the political arena.' Finally, the public sphere is itself subject to the confluence of different codes. In the political-administrative field, investment in people is a major mode of action, while impersonal bureaucratic rules are supposed to govern administrative behaviour. Frank de Zwart's contribution on transfers of government personnel shows the complex effects of merging both ideals. Transfers are meant to keep civil servants from violating bureaucratic norms and to prevent them from getting entangled in local affairs. They are also aimed at an optimal economic use of civil servants by the government.

The effects of the policy are contrary to its official purpose. Though perhaps diminishing parochial corruption, i.e., favouritism dictated by primordial loyalties, it produces 'market corruption';

bribery replaces favouritism. This change reflects the tendency to emphasise impersonal transactions and to minimise unspecified claims by settling debts immediately. However, despite the substitution of diffuse ties by contracts, these seemingly contractual relations remain based on the maxim of investment in people. It creates a setting for continuous efforts to escape the rules by private deals between patrons and clients.

To conclude this introduction the general consequences of the merging of different cultural codes, discussed earlier, will be related to the management of development projects. This will be done by linking the different projects to the framework for the analysis of management processes developed by Moulik and Chaudhury [1986], who distinguish three major management aspects of strategic importance

1. The strategy underlying projects.
2. Processes of strategy development.
3. Processes of strategy implementation.

The strategy underlying projects includes the ideology of the programme and the design of the services.

The ideological aim to improve the socio-economic conditions of the rural poor is central to all programmes concerned. The reports by Ahuja *et al.*, Blankenberg, Satia, Schenk-Sandbergen, Skutsch, Streefkerk, Streefland, and Van der Veen prove how difficult it is to achieve this aim. Structural inequality and its ideological justification are the main obstacles. The chapters by Streefkerk and Van der Veen demonstrate that, as a result of strategic management decisions, programmes even add to the power of already dominant groups.

Involvement of beneficiaries is advocated specifically by Ahuja *et al.*, Satia, and Schenk-Sandbergen as a means of overcoming these hindrances. The awareness and assessment of the differential needs of beneficiaries is a major precondition for this. Not all beneficiaries need the same services, nor do they want them in the same way (Blankenberg, Satia, Schenk-Sandbergen, and Skutsch).

The design of services pertains, among other things, to the nature and role of the implementing organisations. The issue of centralisation or decentralisation of projects falls in this category. It is a major theme in Streefkerk's contribution. Satia and

Blankenberg tackle the same problem with the former mentioning how large projects, unlike smaller ones, may threaten vested interests. Blankenberg argues that centralised planning and decentralised implementation leave little opportunity for local personnel to participate in the design of the strategy of projects; the issue of involvement is raised again.

In the process of strategy development 'influential people' are given a prominent role by Moulik and Chaudhury [1986]. Thus, the analytical framework of management processes includes an aspect that is central to India's cultural perspective. In the contributions by Blankenberg, Satia, and Schenk-Sandbergen, the decisive significance of influential people emerges clearly.

The use of management tools such as needs analysis, generation of alternatives, and the collection of information required for decision-making is another facet of the process of strategy development. Ahuja *et al.*, Blankenberg, Satia, Schenk-Sandbergen, and Streefland show that little information is collected to support decision-making. The scientific assessment of facts is considered unnecessary because authorities believe they already know the facts (Blankenberg). In a cultural context where authority and undisputed responsibility are ascribed to those occupying key positions, the underutilisation of management tools is hardly surprising.

The process of strategy implementation relates to the nature and role of implementing organisations; intra-organisational and inter-organisational coordination; the staffing position of implementing organisations; and the monitoring and evaluation of projects.

Under the heading of the nature and role of implementing organisations, the main subjects are the 'centralisation or decentralisation' issue and the question of 'public versus private', which were discussed earlier.

The role of NGOs is also important here as they serve as intermediaries and implementing organisations in many programmes and provide programmes with local roots. The growing awareness of the importance of projects with a village basis is clear, for example, from the introduction of the Community Health Workers' Scheme in the national health care programme. The tendency to put programmes on a local footing is tied with the cultural perspective that emphasises the cultivation of interpersonal relations. The contributions of Satia, Schenk-Sandbergen, and Skutsch corroborate this view. Skutsch, for example, mentions that programmes are most successful when NGOs have permanent employees to build and repair stoves (chulhas). Blankenberg and Van der Veen argue in favour of engaging private organisations in the implementation of official programmes.

The problem of inter-organisational and intra-organisational coordination and the issue of implementing organisations are closely connected. This is the case when official programmes are implemented by private institutions such as NGOs. Coordination is hampered by factionalism and competition between public and private organisations. These problems are dealt with in the contributions of Blankenberg, and Streefkerk. Competition between public and private institutions, however, is not inevitable. Streefland shows that a sound division of labour can exist between the two. NGOs can play an important role on the periphery as long as official public health organisations are unable to do so.

The staffing of government organisations suffers from frequent transfers (Blankenberg, Streefkerk, and Streefland). Frank de Zwart clarifies how a high turnover of civil servants weakens their morale, lowers their commitment and obstructs a continuous implementation of programmes. Peripheral or backward areas are most affected by the staffing problem; they deserve the best, but get the worst. They are 'penal colonies' for public servants who, for whatever reason, are not wanted elsewhere. Capable people try to 'escape' as soon as possible.

The problem of monitoring and evaluating programmes concludes this review. Almost all contributors indicate the lack of these managerial instruments. In official programmes, the combination of transfers and target approach are detrimental to a proper follow-up. The target approach is the easiest way to judge both results and officials when the turnover of personnel is high. Monitoring and evaluation require a long-term involvement of civil servants.

Monitoring and evaluation is also wanting in NGO projects. Skutsch argues that all parties involved are satisfied by the use of a simple criterion of success. It lessens internal criticism and accusations of favouritism. None of the parties involved will gain by a technically difficult and time-consuming follow-up.

## NOTES

1. See, e.g., Pillai and Baks (eds) [1979].
2. Sanskritisation is the adoption by low castes or tribals of values, norms, rituals and lifestyles characteristic of the so-called 'twice-born' castes, i.e., Brahmin, Kshatriya, or Vaishya.
3. Panchayati raj is a three-tier system (state, district, village) of democratic decentralisation introduced in the early sixties in Gujarat. It was to embody the implementation of national plans on the one hand and local plans on the other. There was an attempt to link the village level panchayat into meaningful larger entities at the district level where operative and administrative power is vested. Under this new set-up the administrative wing was subordinated to the popularly elected political wing. The effect was that the traditionally dominant (caste) groups could continue to exert much power (Panchanadikar and Panchanadikar [1978], pp. 190-216).

# I. HEALTH PROGRAMMES

# 2

# THE DISTRIBUTION OF HEALTH CARE: AN INTRODUCTION

The chapters in this section require an introduction to Indian health care covering the following subjects: the historical background, the distinction between private and official medical care, urban and rural differences, and the condition of children.

In India there are two types of medical care. One comprises the age-old, indigenous methods of health and medical care such as the Ayurvedic and Unani systems. The majority of the practitioners of these systems work in rural areas where they provide medical care to a large segment of the population at low cost, utilising generally non-toxic therapy.

This chapter examines the allopathic or 'modern' medicine introduced in India by the British. The main function of the medical profession was to serve the colonial military and civilian establishments. Before 1857, when the British Crown took over colonial power from the East India Company, there had been no systematic health policy in South Asia. The first Sanitary Commissioners were appointed in 1863. The recurrence of epidemic diseases on an increasing scale, led to a broadening of the scope of the tasks of these Commissioners. Plague was a major scourge which, starting in Bombay, devastated the country between 1896 and 1913. The primary concern of the colonial government was to deal with various epidemics. The Indian Medical Service, staffed by army officers, successfully controlled the incidence of plague, introduced inoculation against smallpox and carried out some useful research in the field of malaria control.

The quality of drinking water was another vital problem. In the First Annual Report of the Sanitary Commissioner for the Government of India in 1864 the following comments were made: 'In many thousands of instances the water of wells (in Bengal) is unfit for human consumption, being loaded with chlorides, nitrites, ammonia and decomposing organic matter; in other words it is continually subject to sewage pollution, and jointly by animal and vegetable impurities of a dangerous character' (Jaggi [1979], p. 174). The health of the soldiers in the British army was a major concern. In 1873, the Commander-in-Chief drew the attention of his officers to the necessity of filtering drinking water. But in many cases filtering appeared to be ineffective because the filters were not clean or did not retain disease producing bacteria. In the meantime, the construction of the first protected waterworks in Calcutta was approved. It was designed to serve 400,000 people and was completed in 1869. Water was pumped from the bank of the river and piped through underground channels to consumers. In 1870, all principal lanes and streets in Calcutta were piped and over 500 streetstandposts were constructed, the total length of piping being 111 miles.

Bombay was the next city to be provided with piped drinking water supply. In 1892, the Viceroy opened the waterworks which were designed to deliver 40 gallons of water per person per day to 525,000 persons.[1] Gradually, piped drinking water became standard in the major cities. Of the total population, however, only a small percentage had access to protected water supply. In Bengal in 1948, for instance, only 7 per cent of the population had access to piped water. In Punjab, 57 per cent of the urban population could rely on piped drinking water, whereas this percentage was less than 1 in rural areas (Jaggi [1979], pp. 174–82).

The introduction of modern education and the consequent rise of a new, indigenous élite was one of the outcomes of British rule. Many members of this élite group opted for the profession of modern medicine so that when the British left India in 1947, medical profession was well entrenched in Indian society (Bose and Desai [1983], pp. 24–26). This is still the case: Indian health care comprises a large private sector and state institutions. The private sector includes traditional practitioners and modern medical establishments consisting of doctors, drug manufacturers, pharmacists, clinics, hospitals and paramedical personnel.

The number of medical colleges increased from 25 in 1947 to 106 in 1984. The annual output of medical graduates grew from 959 in 1947 to 11,992 in 1982 (Bose and Desai [1983], p. 34; Health Information [1986], p. 84).

The state sector has expanded considerably. In 1986, there were 4,093 government hospitals, and 3,381 private hospitals (Health Information [1986], p. 137). It is estimated, however, that in 1978 only 25 to 30 per cent of the doctors was employed in the public sector. And 40 per cent was self-employed and 10 per cent earned a living abroad (Joseph, Desrochers and Kalathil [1983], p. 60).

After independence, the Indian government began to devote its resources more systematically to the improvement of health conditions through Five-Year Plans. In the first five Plans, between 1951 and 1979, the expenditure on the health sector was divided among: (*a*) water supply and sanitation; (*b*) primary health centres, hospitals and dispensaries; (*c*) control of communicable diseases; (*d*) education and training; (*e*) family planning; and (*f*) indigenous systems of medicine.

The largest proportion of expenditure on the health sector in the five Plans was on 'water supply and sanitation' which accounted for 35 per cent in the First Plan and 44 per cent in the Fifth Plan. The bulk of the expenditure on water supply and sanitation was concentrated in urban areas and metropolitan centres. Between 1951 and 1978, the provision of protected water supply in urban areas increased from 50 to 86 per cent of the population. In the villages, the supply of drinking water remained extremely unsatisfactory. In 1980, only 10 per cent of all Indian villages could rely on drinking water from protected sources.

In the period of the Fourth and Fifth Plans, between 1969 and 1979, 'family planning' was the second largest item in the total health expenditure. In the Fourth Plan, it accounted for 26 per cent of the total costs. This figure was 22 per cent in the Fifth Plan.

In order to build up rural primary health care, the establishment of Primary Health Centres (PHCs) and sub-centres was pursued from the Second Plan onwards. In the outlays of the Fifth Plan, the expenditure on these institutional facilities occupied the third place with 19 per cent. In 1980, there were 5,484 PHCs in India, or one PHC per 124,942 inhabitants (Health Information [1986], p. 152).

In 1977, the Community Health Worker Scheme was introduced as an effort to involve local people in government health care systems. The worker provides basic health care facilities to a village or community with a population of 1,000 people. He or she also serves as the first informant on the health status of the community to the medical officer in charge of the relevant PHC.

The Community Health Worker was later called the Community Health Volunteer and more recently the Health Guide (Bose and Desai [1983], p. 38).

The control of communicable diseases like smallpox, malaria, cholera or tuberculosis has also been a major item in successive Plans. In 1974, an international programme was launched against the six major childhood diseases: diphtheria, whooping cough, tetanus, measles, poliomyelitis, and tuberculosis. Another programme discussed in this volume, the Control of Diarrhoeal Diseases Programme, was launched in 1978. The PHCs and Health Guides are involved in the implementation of the two programmes.

The share of the health sector, excluding water supply and sanitation, in the total expenditure varied from 2.9 per cent in the Third Plan to 3.2 per cent in the Fifth Plan (Bose and Desai [1983], p. 32). In the Sixth and Seventh Plans, between 1980 and 1990, this share will be 3.1 per cent (Health Information [1986], p. 70). The share of water supply and sanitation in the total Plan investments increased from almost 1.6 per cent in the Second Plan, to 2.8 per cent in the Fifth Plan and 3.6 per cent in the Sixth and Seventh Plans (Health Information [1986], p. 70).

The health situation in India has improved during the last 40 to 50 years. Large-scale famines have disappeared and the incidence of smallpox, cholera and malaria has decreased considerably. The incidence of malaria has increased again in recent years, but this is a worldwide phenomenon. The life expectancy of an average Indian at birth was only 26 years in 1931; today it is more than 50 (Health Information [1986], p. 52). The crude death rate per 1,000 decreased from 31.2 in 1931 to 12.5 in 1984.[2]

General statistics, however, conceal important differences in the distribution of health services and of mortality patterns between rural and urban areas. The rural population constituted 76.6 per cent of the total population in 1981. These differences are shown in Table 2.1.

The table clearly indicates the urban bias of Indian health care. This conclusion is confirmed by the huge expenditure on urban water supply and sanitation, mentioned earlier, and the high proportion of doctors in urban areas. It is estimated that only 20 to 33 per cent of doctors are found in rural areas (Joseph, Desrochers and Kalathil [1983], p. 60). It must also be noted here that the

TABLE 2.1
**Rural-Urban Distribution of Hospitals, Dispensaries, Crude Death Rate (CDR) and Infant Mortality Rate (IMR)**

| | *Hospitals 1986* | *Dispensaries 1986* | *CDR* 1984* | *IMR* 1984* |
|---|---|---|---|---|
| Rural Areas | 1,567 ( 20%) | 14,463 ( 54%) | 13.7 | 113 |
| Urban Areas | 5,907 ( 80%) | 12,379 ( 46%) | 8.5 | 66 |
| Combined | 7,474 (100%) | 26,842 (100%) | 12.5 | 104 |

**Source:** Health Information [1986].
* per 1,000.

government effort to build up a health care system in the countryside through Primary Health Centres is far from encouraging. Surveys have revealed that rural health centres generally have a poor image and remain underutilised (Joseph, Desrochers and Kalathil [1983], p. 62).

In cities and towns, there are important differences between the affluent sections and the poorest sections of the population. Morbidity and mortality patterns among the urban upper and middle classes approach those in Western countries, whereas the morbidity and mortality patterns of the poorest urban residents are more like those observed in rural areas.

In the villages, diseases arising from poverty, malnutrition, poor sanitation, the unavailability of safe water, the lack of drainage, or of adequate housing, and low levels of immunity are still prevalent. These diseases include TB, gastro-enteritis, malaria, leprosy, filariasis, bronchitis, pneumonia, scabies, worms and fevers (Health [1980]).

The mortality rate of children is high: in 1980, almost one-third of total deaths occurred among children under five (Health [1980]). The infant mortality rate came down from 129 in 1971 to 105 in 1985. Still, it compares poorly with the rates in other Asian countries. Between 1980 and 1984, it was 86.7 in Indonesia, 50.6 in Thailand, and 40.6 in Sri Lanka (Health Information [1986], p. 260). The rural infant mortality rate, which decreased from 138 in 1971 to 113 in 1984, is still very high (Health Information [1986], p. 41).

Health care in Gujarat deviates from the outline presented here which is shown in Tables 2.2 and 2.3.

TABLE 2.2

**Rural-Urban Distribution of Hospitals and Dispensaries in Gujarat in 1986***

| | *Hospitals* | | *Dispensaries* | |
|---|---|---|---|---|
| | *Gujarat* | *All India* | *Gujarat* | *All India* |
| | *(percentage)* | | *(percentage)* | |
| Rural Areas | 8 | 20 | 51 | 54 |
| Urban Areas | 92 | 80 | 49 | 46 |
| Total | 100 (1,236) | 100 | 100 (4,777) | 100 |

Source: Health Information [1986, pp. 135, 145].

* The rural population was 69 per cent of the total population of Gujarat in 1981.

TABLE 2.3

**Rural-Urban Distribution of Estimated Death Rate and Infant Mortality Rate in Gujarat in 1983–84**

| | *Death Rate* | | *Infant Mortality Rate* | |
|---|---|---|---|---|
| | *Gujarat* | *All India* | *Gujarat* | *All India* |
| Rural | 8.2 | 8.5 | 126 | 113 |
| Urban | 12.0 | 13.7 | 56 | 66 |
| Total | 10.8 | 12.5 | 106 | 104 |

Source: Health Information [1986, pp. 37, 38, 42].

The figures in Tables 2.2 and 2.3 indicate a greater urban bias of health care concentration in Gujarat than in India as a whole. The distribution of ownership of hospitals and dispensaries is also revealing. In Gujarat, 70 per cent of the hospitals is in private hands, whereas the majority of Indian hospitals are owned by the government (Health Information [1986], p. 136). More than 80 per cent of the dispensaries in Gujarat is privately owned, the all-India percentage is 42 (Health Information [1986], p. 146).

The very high infant mortality rate in rural Gujarat is the most striking phenomenon which is even puzzling as Gujarat is among the more prosperous Indian states. There are comparatively fewer people in rural areas below the poverty line. For India as a whole this percentage was 40 in 1983, whereas in Gujarat it was 27 per cent (Health Information [1986], p. 267). But the 'below poverty line' criterion hides differences among the poor. The high rural infant mortality rate probably indicates pauperisation among the poorest section of the rural population which is part of the capitalist transformation in Gujarat. Another reason is the high percentage of the tribal population in Gujarat: 14 per cent compared with the all-India figure of 7 per cent.

The close connection between health and economics, i.e., between malnutrition, infection and increased morbidity and mortality and poverty, is well-established (Health [1980], p. 18). Eradication of poverty is the only effective way to improve people's health considerably. This would require a drastic change in the Indian political economy. However, such a change will come about only slowly whereas the health situation demands immediate action.

Programmes that aim to reduce child morbidity and mortality are discussed first in the chapters by Satia and Streefland in this section. This is followed by the official and private treatment of TB. The third theme is the operation and maintenance of drinking water facilities in rural Gujarat. The last contribution examines the public health consequences of the official transfer system.

The theme of Satia's article is how to reduce infant mortality. He first discusses the difficulties of helping more infants to survive. He then reviews three ways of increasing child survival: organisational growth, large-scale programme expansion, and diffusion of innovation. Finally, he discusses the role of NGOs. The personal involvement of political leaders and the mobilisation of all levels of society are the main conditions for improving child survival rates. Satia argues that the government and NGOs should cooperate to achieve this aim.

Streefland deals with the same problem but from an empirical angle. He shows how the general objectives of the Expanded Programme on Immunisation and the Control of Diarrhoeal Diseases Programme are translated into concrete action at the local level. His coverage ranges from international health organisations to villages in North and Central Gujarat. The target achievement system and the personnel management system are the main management systems of both programmes. He explains why the former is most suited for preventive health care such as immunisation. The personnel management system is less adequate as a means of achieving the objectives of the programmes. The power of the medical officer at the district level, who is the personnel manager, is limited because of politics.

Streefland, like Satia, emphasises the limiting effects of

technology on the efficiency and manageability of the programmes. He clarifies the impact of the social, cultural, and geographical contexts on the immunisation and oral rehydration of children programmes. Contextual factors obstruct and shape programmes, while projects, in their turn, reshape their environment. He concludes by saying that NGOs will continue to play an important role 'as long as the government is unable to provide high-level or full coverage by its own health care organisations'.

Van der Veen's contribution highlights the contradiction between public and private treatment for tuberculosis. The article is based on empirical data collected in South Gujarat. The main findings of the study are the poor performance of a district TB centre and the large number of TB patients who patronise private practitioners. This hampers the successful treatment of TB, for either patients cannot afford to pay for prolonged treatment or the doctors do not inform them of the need of lengthy therapy. The patients' preference for private practitioners was noted in the introductory chapter. The cultural code emphasising interpersonal bonds and realising one's own ends through others explains the misplaced confidence in private doctors.

Central to Streefkerk's article is the issue whether to decentralise or centralise drinking water supplies. A comparison between rich and poor areas shows that the operation and maintenance of the water supply in rich villages is not necessarily better than in poor villages in peripheral regions. Centralised systems are found in poor areas and village-based facilities are found in prosperous regions. Proper operation and maintenance of waterworks in villages is obstructed by the intervention of the village élites. Water supply that is decentralised or operated and maintained by individual villages is more liable to face this problem than are centralised facilities. Centralised systems that provide water to many villages from one or a few distant sources do also suffer from the 'grace and favour' nature of the Indian state. However, this effect on the operation and maintenance of centralised provision seems to be less serious than that of the decentralised systems. Streefkerk argues that the operation and maintenance of public provisions such as drinking water should be the responsibility of administrators less involved in village power systems.

Frank de Zwart's article on the official transfer system and its repercussions on Gujarat public health personnel concludes this section. It also highlights the close relation between the local power systems and government employees. Medical personnel cope with frequent transfers as a result of the favours performed through transfer decisions. The consequences for public health care are poorly motivated personnel, absenteeism and the neglect of patients.

## NOTES

1. The first piped water supply in The Netherlands, completed in Amsterdam in 1854, was constructed by British finance and expertise.
2. Between 1980 and 1984, in Thailand this rate was 8, in Indonesia 14.5 and in Nepal 18.9 (Health Information [1986], p. 260).

# 3

# SCALING-UP FOR CHILD SURVIVAL: KEY ISSUES

**J.K. Satia**

Infant and child mortality continue to remain high in India. At the age of 10, life expectancy in India is about the same as that of the developed countries but at birth it is much lower. A child in Indian society faces heavy odds against his or her survival.

The technologies for increasing child survival are available. Many pilot projects have shown that child survival can be significantly increased at a relatively modest cost. However, progress at the national level is slow. The current Infant Mortality Rate (IMR) is 110 per 1,000 live births though it is declining by about three every year. The child mortality rate is 17 per 1,000 per year. Enormous scaling-up of child survival activities is required as less than 20 per cent of the expectant mothers and children currently receive adequate mother and child health (MCH) care. The availability of oral rehydration therapy for diarrhoea, a major cause of infant mortality, is available to an even smaller proportion of the population. Similarly, even though immunisation has been increasing during the last few years, it still does not adequately cover the more vulnerable sections of the population. Those who are poor are not well-informed and are not able to avail themselves of health facilities.

Scaling-up has different meanings in different contexts. Here the term scaling-up applies to the approaches needed to expand a variety of technologies, activities, and programmes. A few randomly selected examples are:

— Several voluntary organisations developed rural health programmes which had village health workers as their base
— before the government implemented the Village Health Guide Scheme.
— A voluntary agency had been working in a small area. It is now engaged in expanding its coverage to about 100,000 persons.
— The Integrated Child Development Scheme (ICDS) had been tested earlier in pilot blocks before being expanded to cover a larger part of the population.
— A voluntary agency had been providing rural health services in a small area. It has now taken over several health centres that had been operated by the government.
— Gradually the universal immunisation programme is being expanded to cover the whole country.

This chapter discusses the key issues in scaling-up[1] by first identifying the constraints. It then discusses three common modes for scaling-up: organisational growth, large-scale programme expansion and diffusion of innovations. The key issues in scaling-up are identified as mobilisation of demand, securing adequate financial and human resources, selecting an appropriate mix of activities, modifying structures to match strategies, training staff, and instituting a management information system and operational research. The relative emphasis on these issues will depend upon the specific modes used for scaling-up. The paper concludes with a brief discussion of the role of non-governmental organisations (NGOs) in scaling-up for child survival.

## 3.1 CONSTRAINTS TO RAPID SCALING-UP

A review of past experiences suggests that scaling-up is not an autonomous process and, more often than not, even scaled-up activities as a result of external stimulus are not as effective as they are during the pilot stage. The need for effective scaling-up has been felt for quite sometime now but difficulties in achieving this effectively have also been a cause of concern to all those involved—usually NGOs, public sector programme managers, domestic fund providers such as planning and financial ministries, and foreign financial agencies. But an examination of the literature suggests that insights into the processes of successful scaling-up are inadequate.

An analysis of the efforts to scale-up health and other sectors of the economy, such as the Green Revolution, family planning programmes, literacy campaigns and immunisation campaigns has led to the identification of three conditions for nationwide coverage (Mandl [1984], p. III): (*a*) personal involvement of political leaders, (*b*) mobilisation of all levels of society, and (*c*) promotion of every available means to deliver new knowledge and a simple technology to those whose lives can be improved by it. These conditions imply the type and amount of effort necessary. But political will and community resources need to be harnessed to bring about national coverage. The constraints to a rapid scaling-up, therefore, are likely to be lack of availability of an appropriate technology, adequate resources, organisational or institutional capabilities, and necessary managerial skills.

### 3.1.1 Technology for increasing child survival

Although the technologies for increasing child survival are available, they have certain inherent limitations to being scaled-up. For instance, most of the vaccines are not heat resistant and cold chain (see chapter by Streefland) is difficult to maintain in most developing countries. This limitation has constrained scaling-up of immunisation services. Since a programme has to depend upon the existing technology, the question is how these technologies can be refined, made more widely available, and affordable. The requirements of each technology also differ. For example, the technological capability required to scale-up immunisation is higher than that required to scale-up appropriate breast-feeding practices. The requirements of the same technology also differ from place to place, which need to be considered along with organisational factors or procedures when transferring learnt experience from one setting to another.

### 3.1.2 Availability of resources

The resources allocated are usually less than those required for a desired level of scaling-up for child survival. The availability of resources also, in turn, determines the availability of technologies.

For instance, the availability of vehicles, equipment and other hardware can ensure cold chain. Therefore, it is essential that the government provides sufficient financial resources. Nevertheless, a lack of adequate resources may not be a serious constraint in scaling-up since the amounts required are not large. Second, the resources allocated to these activities have been increasing over time. Finally, the community itself can provide most of the physical resources required for these activities.

### 3.1.3 Organisational capabilities

Financial resources alone may not suffice. Considerable emphasis needs to be placed on the structural organisation and human resources. The organisations need to develop capabilities for generating support for child survival, for implementing scaled-up activities and for harnessing community resources.

### 3.1.4 Necessary managerial skills

As the scale of the activities increases and as they become diversified, the skills required to manage them efficiently and effectively also need to be developed. Often the management aspect is neglected and the lack of necessary managerial skills poses a constraint.

## 3.2 MODES OF SCALING-UP

Although it is difficult to develop a taxonomy of approaches, three basic modes of scaling-up can be distinguished: organisational growth; large-scale programme expansion for provision of products and services on a wider scale, usually by replication of a pilot project; and diffusion of innovation to expand the use of desired practices. Each of these modes are appropriate for certain technologies and contexts. However, several managerial issues arise with respect to each of them.

### 3.2.1 Organisational growth

In this mode the organisation grows, either by providing a wider range of services, or covering a wider area, or both. In its initial years, a voluntary organisation, or a programme looks for appropriate services, activities and organisational structure, and its most important concern is to establish credibility with the people, among peers, and other social institutions. Tasks are performed as a team effort, the structure is informal, practices are evolved, and considerable experimentation in services takes place. The financial resources are meagre and pioneers require a high degree of commitment to face many early hardships (Subramanian [1983]).

If successful in its early stage, the next phase could involve consolidation in which some definite pattern of work emerges based on acquired experience; the team continues to function but role descriptions and distinction among members of the team become more clearly defined. Some practices and systems become established and the 'credibility earned' brings about resource stability.

Once proven, internal and external demands for expansion are made. Sometimes, such demands are resisted by the early pioneers because size inevitably demands more formalism and standardisation and an administrative structure to go along with it. Decisions taken at this stage may also distort the identity of the organisation and may dilute its value system. The managerial demands on the organisation also increase and functional specialisation is needed.[2]

### 3.2.2 Large-scale programme expansion

This mode of scaling-up is frequently used to develop large programmes from small pilot projects. Most successful programmes, indeed, are the results of learning from pilot projects. The main concern, however, is that the programmes do not perform as well as the initial pilot project.

Researchers have paid considerable attention to the issue of loss of effectiveness during project expansion (Pyle [1984]). There are several reasons: first, considerable deviations seem to occur during implementation; goals may not be set appropriately, hierarchical bureaucracies set up for implementation may not be responsive to local needs and may not make necessary adaptations; demand creation activities may receive less emphasis and staff may not be as motivated while support in supervision and logistics may be inadequate. Second, while a small project may not threaten the existing social, economic, political and professional interests, a large project may do so. While the leader of the small project may be able to interact with politicians and bureaucrats this may be neglected during expansion. Finally, because of resource exigencies, only some elements of the project may be adopted for expansion while other elements may be ignored; the synergy among various elements, therefore, may be lost to a large degree.

In summary, the experiment-implementation-performance gap could be largely attributed to a failure of strategic management (Paul [1982]). The strategies may not be suitable for the environment; while the programme processes of the organisation may not match the strategies selected.

### 3.2.3 Diffusion of innovation

In this mode, the use of a specific practice, which may include the use of specific products or services, is scaled-up. The process is often characterised as consisting of two parts: adoption and diffusion. During the initial stages, a few people adopt a practice. These early adopters are usually willing to accept changes, and take risks associated with the use of new practices. Growth is slow to moderate during this phase. However, once a critical threshold level of usage is reached, diffusion occurs. During this phase, there is sufficient autonomous communication within society and the rate of increase in practice is high until it reaches a saturation point.

Two major activities required during diffusion are dissemination of information and demonstration. The purpose of dissemination is to create widespread awareness of the innovation among practitioners. The criteria which are to be used in evaluating dissemination activities include intelligibility, accuracy, pervasiveness and message impact. The objective of demonstration is to provide an opportunity to examine and assess the operating qualities of the innovation. Demonstration can be evaluated in terms of its credibility, convenience and evidence assessment.

In addition to effective dissemination and demonstration, other activities may be used to aid the process of diffusion. These activities include helping the practitioner by providing support in the form of advising or trouble shooting; involving practitioners either individually or as a group in the process of problem identification; developing, testing, and packaging services; seeking the active help of practitioners in diffusing the services to others; and training them in the use of the innovation.

In the available literature on child survival, social marketing as a mode of scaling-up has not received adequate attention (Ramachandran [1985]), primarily because market mechanisms are not expected to work in the area. However, a large number of services and activities directed towards increasing child survival are provided through commercial channels, particularly in urban areas. Therefore, conscious attempts need to be made in this regard.

#### 3.2.4 Technology and the mode of scaling-up

The mode used should be appropriate for child survival activities to be scaled-up. Diffusion is appropriate where practices such as oral rehydration therapy (ORT), breast-feeding, safe delivery, or growth monitoring are concerned. Commercial channels can be used through social marketing for distribution of oral rehydration salt (ORS) and contraceptives. Government programmes are perhaps the most appropriate way to provide services such as MCH care and campaigns to promote nutrition. In the early stages, and in underserved areas, NGOs may be able to provide these services.

### 3.3 ISSUES IN SCALING-UP

Issues in each of the above modes will differ. A review of the literature and of selected successful experiences suggests that the key issues are mobilisation of demand, securing adequate financial and human resources, selecting an appropriate mix of activities, modifying the structure to match strategies, training staff, instituting a management information system, and operational research. The emphasis on these issues will depend upon the specific modes used for scaling-up.

#### 3.3.1 Mobilisation of demand

If the demand already exists, it is easier to scale-up which often occurs by expansion of service delivery. However, the major problem in scaling-up is the lack of demand. Therefore, careful attention to demand creation is necessary.

Aroles clearly emphasises demand led expansion in a rural health project in Jamkhed (Satia *et al.* [1985]). A village is not added to that programme unless it so requests and is willing to extend the necessary cooperation. However, steps were taken to create such a demand. The entry point sought was based upon the felt needs of the community mainly for curative services. Organisations of relevant groups of farmers, women and youth were formed to create and sustain demand.

Through the use of such groups the community also sets an example for motivating other communities and thus aids in scaling-up. While this approach ensures demand and makes scaling-up easier, it suffers from an obvious dilemma: communities most in need will perhaps not be the first to request services and the need for services which are most desirable for increasing child survival may not be felt at all.

A related question is that of segmentation of clients. All clients neither require the same services nor do they want the services in the same manner. It is generally agreed that market mechanisms should be used to meet the needs of those who can afford to buy them. The question arises as to whether NGOs should serve the middle class who can pay at least for part of the services or should they serve the poor who cannot afford to pay at all. In the latter case, NGOs will be dependent on the government for financial resources because only the government can supply resources on a sustained and continuing basis.

Community processes have an important role to play in creating demand. The degree of community involvement may vary: the community may just be aware of the activity, it may be mobilised for a specific purpose or it may participate in an ongoing programme. Achieving such participation is likely to be a difficult process. A major initiative of the government to secure community participation in health programmes was the Village Health Guide Scheme. Most observers suggest that these Health Guides do not serve as change agents; the responsibility of providing various services is assigned to them.

Involving the community means involving the most unskilled persons with low levels of literacy. The educational efforts implied in this process are hence extremely time-consuming and expensive.

### 3.3.2 Securing financial resources

How to overcome constraints of financial resources for organisational growth? The answer is: by charging for curative services which people need and demand. The rural health project in

Jamkhed has become self-reliant in terms of its operating costs but in the initial stages external funding was necessary. The scale of operations has to be large enough to generate sufficient resources for this purpose and to benefit from economies of scale. To save money, the government does not provide enough resources for support services such as training, supplies, research and evaluation. Organisations such as the National Dairy Development Board (NDDB) have used innovative strategies (Paul [1982]). The financial resources used to expand Indian dairy activities through Operation Flood-I were generated by selling milk prepared from milk powder made available through foreign aid. This secured financial autonomy from the government. The strategies for generating resources must be such that they do not inhibit organisational growth. A lack of financial resources makes it difficult for the Jamkhed project, for example, to attract the desired quality of people.

### 3.3.3 Securing human resources

Another important issue is the development of human resources and the organisational process needed to create the appropriate value system. The NDDB in India employed spearhead teams to establish village level co-operatives. Recruitment, training and staff motivation were undertaken to create the necessary missionary zeal and commitment to the programme among staff workers. Once again, there cannot be a single answer to this problem. Carefully worked out strategies need to be adopted if a lack of the required quantity and quality of human resources is not to constrain seriously organisational growth.

### 3.3.4 Modifying structure

Often the limitations of human or financial resources are those imposed by the organisational structure. A functional organisation usually constrains scaling-up whereas a divisionalised structure promotes growth. In Operation Flood-II, the NDDB attempted to create organisations which can become independent and self-reliant in all respects. Many NGOs are able to expand because they set up separate divisions for their diversified activities (Tanothai [1985]).

For the success of programmes, such as the Integrated Child Development Scheme (ICDS), which involve health and nutrition services, two points are important: coordination between the

ICDS and the health programme, and teamwork of the three workers at the village level, namely, Anganwadi worker (kindergarten-cum-health worker), Village Health Guide, and Traditional Birth Attendant (Contractor [1985]). The job responsibilities of the three health workers are different but complementary. A variety of formal mechanisms has been set up to coordinate the work which extends all the way up to the cabinet level. The medical officers are also given orientation training, but coordination continues to remain weak.

### 3.3.5 Selecting an appropriate mix of activities

There are many factors affecting child mortality. Scaling-up activities which may have a significant impact on child mortality is not enough. It is more important to talk in terms of scaling-up the impact of activities on child survival rather than only of the activities themselves. The package of services to be scaled-up should be specifically defined after studying the problem of child survival in a particular area and analysing and evaluating available interventions to improve child survival. Since the determinants of child mortality vary considerably from place to place, epidemiological studies should be undertaken before such a package can be appropriately defined, especially for large geographic areas or at the national level. A pilot project can be undertaken to assess the impact of implementing this service package.

### 3.3.6 Instituting a management information system

Almost all programmes and organisations attempt to develop monitoring of more comprehensive management information systems (MIS), but the managers still do not get the feedback quickly enough. As the consumers do not directly pay for the services, the managers may not be able to tune the delivery system to their needs. The problem is compounded because the impact is difficult to evaluate. To ensure that the impact, effectiveness and quality are retained when scaling-up takes place, a good MIS is necessary.

### 3.3.7 Using operational research

Often programmes are launched without an adequate pilot study. Sometimes, NGOs carry out the pilot projects and the government implements them on a larger scale. Quite often when an innovation is scaled-up, it is the philosophy and not the process responsible for the success of the innovation that is actually scaled-up. Such processes should be considered carefully when planning for scaling-up. Most experiments are initially not designed to be scaled-up and is generally considered only after a programme achieves success. There is a better chance of success if an experiment is designed with the objective of scaling-up. If this is not the case, the various processes that contributed to the success of the original experiment should be carefully examined before it is considered for large-scale expansion.

### 3.3.8 Training staff

Often under pressure to expand rapidly, training of staff is neglected. Training is perceived very narrowly as the task of providing skills and on-the-job training. Work habit formation and continuing staff development do not receive adequate attention. In expanding the ICDS programme, similar difficulties were experienced. While the quality of training was good in the initial stages, at the later stage there were no training institutions of good quality and training suffered which then affected the performance in the field.

### 3.3.9 Strategic shift

There are also the dynamics of scaling-up to be considered. At one level of achievement the programme may be highly successful. As the level of achievement increases, the programme needs to be modified and to remain successful a strategic shift has to occur. Such a sequence has been well documented for the Indonesian Family Planning Programme (Basic [1982]). The programme started by obtaining commitment from religious leaders and by operating mobile service teams. As the level of contraceptive coverage increased to about 15 per cent in an area, village posts were established. When the coverage reached about 35 per cent in a village, acceptor groups in the form of mothers' groups were formed. As the coverage continued to increase, the programme added growth monitoring and nutrition schemes. At present experiments are being conducted with an income generation scheme. Although the basic structure has remained the same, it has been adapted to new strategies about once every five years.

## 3.4 ROLE OF NGOs

Few NGOs operate on a national scale, most of them are localised. The contribution of NGOs to child survival could be improved by either increasing the number of NGOs or helping them to broaden their service coverage, if they are willing to expand. NGOs could be involved as partners of the government in the provision of services, particularly for underserved areas. They could also fulfil specific tasks such as training which is their major strength. They could build a supportive environment by creating a demand for services and motivating people to accept and utilise the services provided. An organised NGO sector will facilitate policy processes by putting pressure on the government to ensure better programme performance.

The collaboration between NGOs and the government will be more effective if there is more interaction at all levels between the two. However, the NGO should be able to distinguish between its role as an actor and as a catalyst. First, it should realise the difference between the roles, and second, it should know how to perform that role. The NGO should be careful while taking over government programmes because it may not have sufficient administrative experience of managing large systems and it may not wish to involve itself in routine matters. Therefore, as catalysts, NGOs do not need administrative control but should provide technical advice.

To scale-up child survival activities necessitates carefully thought out strategies of involving NGOs and a clear delineation of their role. This role will also change as scaling-up takes place. In the initial stages NGOs could act as innovators. During the expansion stage, they should support government efforts and act as catalysts. As the coverage improves, NGOs should be active in demand generation among communities where such coverage remains poor. It is a challenge that the government and NGOs should meet as partners if effective and rapid scaling-up for child survival is to take place.

## NOTES

1. In August 1985, the Indian Institute of Management, Ahmedabad organised a regional workshop on Scaling-up for Child Survival activities which was sponsored by the Ford Foundation. The objectives of the workshop were to share experiences, to identify key issues and to develop conceptual and practical approaches for the problem of scaling-up. This paper is partially based on some of the insights derived from the workshop.
2. These phases are somewhat similar to the three stages mentioned in Korten [1980, pp. 480–511] as learning to be effective, learning to be efficient, and learning to expand.

# 4

# MANAGEMENT AND UTILISATION OF IMMUNISATION AND ORAL REHYDRATION THERAPY IN RURAL GUJARAT[1]

**Pieter H. Streefland**

Development of more heat-stable vaccines and large-scale use of scientifically approved, home-made oral rehydration mixtures form the technical basis for a potentially massive reduction in infant and child mortality. In the development, testing and diffusion of these new technologies, international agencies, like UNICEF and WHO, large research institutes, like ICDDR (B) and CDC, and consultancy firms, such as John Snow Inc., have played a crucial role.[2] For them, application of these technologies has acquired world-wide dimensions. They have formulated strategies for implementation by national, regional, and local public health authorities.

The new technologies are embodied in two world-wide programmes—the Expanded Programme on Immunisation (EPI) and the Control of Diarrhoeal Diseases Programme (CDD). Their objectives and modes of operation are defined at high administrative levels. Their implementation is largely a local level affair, concerning the activities of several middle level public health staff, coordinators, trainers, technicians, and Village Health Workers, who have to work concertedly to make the programmes effective. Their work is carried out in a specific social, cultural and geographic environment. This includes the social organisation of village society, settlement and migration patterns, and ideas and beliefs regarding prevention and cure of the diseases concerned. This chapter tries to relate the conception and management of the EPI and CDD programmes at the international, national and regional levels to their implementation at the local level in the specific context of rural Gujarat. It is an effort to trace the process of realisation of general concepts, objectives and technical possibilities. The emphasis will be on the final shaping of both programmes at the local level i.e., in the villages of Gujarat.

The study focuses on the following issues: (*a*) the processes through which the general objectives of the EPI and CDD programmes have been given concrete form at the local level; (*b*) the management tools that have been selected (and the ones that have been omitted or neglected) to attain the objectives in Gujarat and how they are applied; and (*c*) the influence of the environment on the results of the programmes.

The study began with a review of literature on immunisation and oral rehydration therapy. This was followed by two brief periods of fieldwork. In the third quarter of 1984, general information on the operation of the public health services in Gujarat and, more specifically, on the status of the EPI and CDD programmes in the region was collected. For this purpose extensive interviews were held in Ahmedabad, Nadiad and several Primary Health Centres (PHCs) in Kheda district. In mid-1985 the author returned to Gujarat for a period of five weeks. An intensive programme of field visits was organised both in Kheda and in Banaskantha districts. Interviews were held with public health staff, Village Health Workers and village women. Ongoing immunisation work was also observed.

## 4.1 THE PROGRAMMES AND THE STRUCTURE OF PUBLIC HEALTH CARE

Immunisation against six major childhood diseases, namely, diphtheria, pertussis (whooping cough), tetanus, measles, poliomyelitis and tuberculosis, is at present being strongly advocated by the EPI of the WHO, and by UNICEF in its selective primary health care approach known as GOBI.[3] The EPI was launched in 1974 following the success of the smallpox eradication programme. The CDD programme, aimed at reduction of infant mortality resulting from diarrhoea, is a more recent development (May 1978). As in the case of immunisation, both WHO and UNICEF support the CDD programme by way of donations, developing training materials and organising training courses, and also by exerting pressure on national public health care authorities.

The EPI and CDD programmes are very dissimilar, reflecting the characteristics of the technologies they embody. This dissimilarity, outlined in Table 4.1, was the main reason for selecting these two programmes for the study, the intention being to gain additional insights by contrasting them. At the central government level in Delhi, the EPI and CDD programmes are clearly demarcated organisational entities in the Ministry of Health and Family Welfare. The programmes were organised by the national government in 1978 and 1984, respectively.

At the state level, the programmes lose some of their distinctness as they become intertwined with other forms of health care. The essential integrating mechanism is the concentration of responsibility for a number of different health care functions in the hands of the same personnel. Formerly, there used to be specialised workers, such as malaria workers or family planning workers. In 1975 this principle of specialisation was abandoned and a new category of public health staff was created, aptly called multipurpose health workers (MPW). At Gujarat state level, a joint director and a deputy director, both under the Director of Health, Medical Services and Medical Education are responsible for the EPI and CDD programmes respectively.

The district health administration is headed by the District Health Officer (DHO) (in Kheda district there is a Chief Medical Officer of Health), assisted by administrative and supervisory staff, and by medical officers with specific responsibilities in the fields of epidemiology, family planning and communicable diseases. Public health care is provided by PHCs (each covering 80,000 to 100,000 people), male and female MPWs (each covering 5,000 to 8,000 people) and by Health Guides (HG) who work at the village level.[4] The EPI is implemented by male and female MPWs who are supervised by the PHC and supported by the HG. In oral rehydration therapy (ORT) the division of labour between MPWs and HGs is not clearly demarcated; supervision comes, again, from the PHC.

### 4.1.1 Implementation of the programmes: a crude picture

The cycle of public health activities is planned on a yearly basis from April to March. During a particular year, the MPWs have to carry out a variety of activities in the working area allotted to them. The size of the population they cover may vary considerably, depending on the density, the settlement pattern, and vacancies in the surrounding areas. In family planning (FP) and immunisation programmes, MPWs often cooperate with colleagues to form teams. Generally, vaccination camps are held during July, August, and September, when summer is over and schools have reopened. This yearly campaign, comprising at least three camps, begins with a survey of children and pregnant women eligible for vaccination. Next, the team visits a village and vaccinates the listed persons, aiming to reach the targeted number for that year. Such a team basically includes a supervisor from the PHC, a male and a female MPW and the local HG. Larger and smaller teams were observed as well. Usually an immunisation session is held in someone's house. A team always tries to immunise as many eligible individuals as possible in one place. This saves time, as they are able to use the same boiling water for sterilising instruments for all the immunisations. According to the rules, village women should be informed in advance about the visit by a team. Dates are fixed every month at the PHC and transmitted to the village by the visiting health staff or by the HG. But this is a weak link as all villages do not have an HG. Moreover, MPWs visit them only once a fortnight.

Immunisation sessions are also used for providing health education, for instance, on family planning for mothers. For carrying out the main duties of the team, such as bringing together children and pregnant women, record-keeping, immunisation and the sterilisation of instruments, two skilled workers suffice, but they take the support of the HG in assembling the people. The HG benefits in terms of the prestige gained from being seen working with highly trained health staff. In all cases, the public health staff appeared to be experienced and efficient in their work. Children who are ill or unable to attend a camp and therefore miss one or more doses, are vaccinated by MPWs during their routine visits to the households. These routine vaccinations can be given any time during the year provided the vaccines are available from the PHC,

TABLE 4.1

**Comparison of Immunisation and Oral Rehydration Technologies**

| | *Immunisation* | *Oral Rehydration* |
|---|---|---|
| 1. Objective: | Reduction of infant and child mortality due to six childhood diseases | Reduction of infant mortality due to diarrhoea |
| 2. Means: | Vaccination to boost the immune response of the body (preventive) | Oral rehydration in case of acute diarrhoea to combat and prevent dehydration of the body (curative; case management) |
| 3. Rules for applications: | Many strict rules, regarding:<br>cold chain maintenance<br>vaccination age<br>number of doses and their time sequence<br>proper vaccination technique<br>calculation of targets | Few strict rules regarding the mixture of the ingredients of the solution (salt, sugar or jaggery, boiled water or the contents of a prefabricated ORS packet with boiled water) |
| 4. Control: | The rules to be followed primarily by public health staff, as they control the application of the technology | The rules to be followed primarily by mothers or guardians of sick infants, as they control the application of the technology |
| 5. Expertise required: | Depending on the organisational level:<br>vaccination skills<br>management skills<br>monitoring skills | Primarily skills of making the mixture and the recognition of the need for application |

| | | |
|---|---|---|
| 6. Know-how to be transferred to the target group: | Mothers (guardians) have to understand the importance of prevention by vaccination, as well as the need for an infant to have several vaccinations. Moreover, they have to be told about the possible side-effects | Mothers (guardians) have to understand about the danger of dehydration and the importance of rehydration. Moreover, know-how regarding preparation of the solution has to be transferred |
| 7. Outside connection: | The technology (vaccines, vaccination and cold chain equipment) is heavily dependent on supply from outside | The technology does not require any outside supply, though the option of supply of ORS packets is there |
| 8. Possibilities for planning and organisation: | As the timing of the vaccinations is important and, from a medical point of view, rather independent of seasonality, as the aim is to immunise many children in a certain age group at any one time and as the expertise is controlled by the public health staff, this technology is extremely suitable for a strictly planned, campaign-like approach carried out by a hierarchically structured organisation | As the administration of ORS is dependent on individual cases of acute diarrhoea (incidence of which will probably show seasonal fluctuations), and as the expertise is largely controlled by the target group and there is no necessity for outside technical dependence, this technology is most appropriate for a participatory approach with emphasis on transfer of knowledge. The *ad hoc* character of treatment further stresses individual responsibility of mothers and public health workers |

where they are stored under prescribed physical conditions. School-age children are vaccinated by teams which provide health education during the same visit. Immunising older children is not an easy job. We once saw a whole school deserted within a few minutes after a lookout had announced the imminent arrival of the team.

Both routine and camp immunisations are performed according to the immunisation schedule and in order to reach targets. Immunisation schedules and targets are decided for the executive workers from the top. The schedule (Table 4.2) is fixed at national and state levels following international specifications. The targets

TABLE 4.2
**Immunisation Schedule Followed in Gujarat**

| *A. Children* | *Vaccination* |
|---|---|
| Age of Children: | |
| 3–9 months | BCG vaccine (one dose) |
| | Triple vaccine (DPT) (three doses with at least four weeks between each dose) |
| | Polio vaccine (three doses with intervals of at least one month) |
| 9–12 months | Measles vaccine (one dose) (in rural areas practically not available) |
| 15–24 months | DPT (booster dose) |
| | Polio (booster dose) |
| School Age: | |
| 5–7 years | Bivalent vaccine (DT) (booster dose or two doses if not vaccinated previously) |
| | Typhoid (two doses) |
| 10 years | TT (against tetanus) (booster dose or two doses if not vaccinated previously) |
| | Typhoid (booster dose) |
| 16 years | TT (booster dose) |
| | Typhoid (booster dose) |
| *B. Pregnant Women* | |
| Duration of pregnancy: | |
| 20–26 weeks | TT (first dose) |
| 30–36 weeks | TT (second dose) |

**Source:** A.A. Contractor and M.H. Shah [1982b, p. 9] (with slight adaptations).

are determined at each organisational level for the level immediately below, using guidelines formulated at the international, national and state levels. The targets are expressed in terms of the number of children and pregnant women to be covered for the number of doses of a specific vaccination. Targets are specified for the whole country, for each state, district, PHC, and individual MPW. The size of the target is based on the number of people eligible in each age category which can be estimated rather simply from the demographic data: the total population, to be extrapolated from the 1981 Census; birth rate; infant mortality rate; and age specific death rate. When the number of eligible people is known, the target population is fixed as a certain percentage of this figure. The state level targets for Gujarat fixed in Ahmedabad (Table 4.3)

TABLE 4.3
**Annual Coverage of Targets for Specific Immunisations in Gujarat**

| *Year* | *DPT** | *TT*** | *Polio** | *DT**** |
|---|---|---|---|---|
| 1974–75 | 400,000 | 290,000 | — | 400,000 |
| 1975–76 | 400,000 | 290,000 | — | 400,000 |
| 1976–77 | 700,000 | 400,000 | — | 400,000 |
| 1977–78 | 700,000 | 600,000 | — | 600,000 |
| 1978–79 | 700,000 | 600,000 | — | 700,000 |
| 1979–80 | 2,000,000 | 500,000 | — | 2,000,000 |
| 1980–81 | 1,200,000 | 600,000 | 16,000 | 1,500,000 |
| 1981–82 | 1,200,000 | 1,000,000 | 196,000 | 1,500,000 |
| 1982–83 | 1,000,000 | 800,000 | 250,000 | 1,200,000 |
| 1983–84 | 1,000,000 | 900,000 | 600,000 | 900,000 |

**Source:** Maternal and Child Health Programme Information, Gujarat, Ahmedabad, 1983.

* Course is three doses; ** course is two doses; *** booster dose.

tend to be a bit on the higher side than those handed down from the Centre. The state administration is of the opinion that in this way, the chances of achieving the targets of the centre are better. This practice accounts for some of the fluctuations in targets over time as shown in Table 4.3.

The availability of vaccines and maintenance of the cold chain are very crucial. At present, all vaccines, except polio and measles are manufactured in India. Polio vaccine is imported in bulk and then diluted and bottled in Bombay. Vaccination against measles has only recently been introduced in the Indian EPI. This vaccine is very expensive and requires a very reliable cold chain. At present only small quantities are needed, as vaccination against measles is, as yet, mainly restricted to urban areas. The lack of a sufficient supply of vaccine was mentioned as an important factor greatly hampering immunisation performance. With regard to polio vaccine, for instance, during 1984–85 Gujarat received only 75 per cent of the vaccine needed, while for 1985–86 only half of the quantity needed was expected.

The cold chain is a system of transporting and storing vaccines at recommended temperatures from the manufacturer to the point of use. Its purpose is to guarantee the efficacy of the vaccines.[5] Generally, vaccines must be stored and transported between +4° and +8° centigrade; polio and measles vaccines can be kept at sub-zero temperatures, whereas DPT, DT, TT and typhoid vaccines should not be frozen. Vaccines should not be exposed to direct sunlight. Polio and measles vaccines are damaged very easily. Besides refrigerating equipment, the cold chain material at the sub-state levels comprises vaccine carriers with ice packs and thermo-cool boxes which are damaged easily and are not easily replaced. The limitations of the cold chain make team vaccinations preferable to immunisation by a single public health worker. The need for boiling water to sterilise instruments is another technical snag which makes vaccination camps preferable as the same pot of boiling water can be used for a larger number of immunisations.

The CDD programme is more recent and less developed than the EPI programme, although diarrhoeal diseases are stated to be one of the major causes of childhood morbidity and mortality in India. The international CDD programme has four pillars: epidemic control, maternal and child health care practices, environmental health practices and case management. In the CDD programme in India and in Gujarat the emphasis is at present on the last pillar, one which is directly related to the short-term objective of reducing deaths from diarrhoeal diseases. General targets have been formulated, for instance, 35 per cent, 75 per cent and 95 per cent reductions in 1985, 1990 and 2000 respectively.[6] It is, however, extremely difficult to monitor progress towards achievement of these targets as this requires field surveys, including a good base-line study, for which the public health services are presently not equipped.

The major means adopted to accomplish this short-term goal is

ORT. For its implementation the public health staff at all levels have been familiarised with the assumptions and rules of the therapy. They are provided with pre-fabricated packets of oral rehydration salts (ORS), and are also taught the home-made mixture of salt, sugar or jaggery, and boiled water. The programme has no clear preference for either of the options. The packets do look more professional, but are often in short supply. Ideally, when advocating oral rehydration, the public health staff should also pass on other health information, like the need for continuation of breast-feeding during the diarrhoea.

Though there are many diarrhoeal diseases, the MPWs and HGs basically have to recognise only acute diarrhoea and prevent death due to dehydration. Moreover, they should be able to decide when their knowledge is too limited to assist, and the child should be referred to a more sophisticated source of health care. In practice, the matter is a bit more complicated, as the public health staff are provided with other curative means against diarrhoea as well. The MPWs often have at their disposal a binding agent, a drug against bacterial dysentery, and another against amoebic dysentery, in addition to ORS. The binding agent is also part of the HGs' stock. Hence, when treating a case of diarrhoea, the public health worker may be confronted with a variety of options, the exact number depending on the actual drugs in his or her possession.

## 4.2 THE STUDY AREAS

In order to obtain sufficient information, an exploratory, comparative study was designed. One central district (Kheda) and one peripheral district (Banaskantha) were selected for data collection to enable comparison of problems faced in a relatively more developed area with a less developed location. In each of these districts a further selection was made of two PHC areas, one central and one peripheral. In Banaskantha, Varahi and Chandisar PHCs were selected, and in Kheda, Alindra and Birpur PHCs were selected.

An important condition for the quality and the coverage of the EPI and CDD programmes is the personnel situation, especially at the levels of MPWs and HGs. The way in which the four selected PHCs compare in this respect is presented in Tables 4.4 and 4.5. The difference between the two PHCs in Kheda regarding the

TABLE 4.4
**MPW Posts and Vacancies under Selected PHCs (1 September 1985)**

| *PHC* | *MPW Posts Allotted* | *Vacancies* | |
|---|---|---|---|
| | | *No.* | *%* |
| Varahi | 29 | 6 | 20.7 |
| Chandisar | 40 | 8 | 20.0 |
| Birpur | 26 | 7 | 26.9 |
| Alindra | 51 | 6 | 11.8 |

**Source:** Interviews at PHCs.

TABLE 4.5
**Health Guides Working under Selected PHCs**

| *Year* | *Number of HGs Working** | | | |
|---|---|---|---|---|
| | *Varahi* | *Chandisar* | *Alindra* | *Birpur* |
| 1979–80 | 56 | 64 | 108 | 100 |
| 1980–81 | 56 | 87 | unknown | 96 |
| 1981–82 | 56 | 87 | unknown | 86 |
| 1982–83 | 56 | 87 | unknown | 106 |
| 1983–84 | 46 | 95 | unknown | 98 |
| 1984–85 | 36 | 95 | 100 | 94 |

**Source:** Interviews at PHCs.
* In Alindra, Chandisar and Birpur a new batch of HGs was trained in 1980. In Varahi no new candidates could probably be found.

percentage of MPW vacancies though striking, is understandable, as Alindra is well connected with Nadiad, the district centre and includes several well-to-do and large villages. Birpur PHC is far away from Nadiad, and covers the forest area and villages which allegedly harbour criminal elements. Punishment is often the major reason for posting the PHC staff to this area. The fringes of the Birpur area are considered unsafe for female MPWs. Varahi PHC area is also at a distance from the district centre, but it is situated on the main road to the Saurashtra region from where many MPWs come.

Another factor, namely, the relative perceptions of different areas, should be mentioned here. To people working in Ahmedabad and in other districts, Banaskantha is a backward area which some even view as a place to be transferred to as punishment. Those who work in Banaskantha usually share this negative view of the district although Varahi is not considered to be very peculiar, as there are more remote places in Banaskantha. Birpur, on the other hand, is considered much more of an exception in the district, a black spot in a central and prosperous region. To observers in Ahmedabad and for those who live and work in Kheda district, the Birpur area is extremely peripheral. Accordingly, the feeling of relative deprivation among the Birpur health staff is considerable.

Regarding the availability of HGs, the data are not fully comparable. Some data for Alindra PHC could not be obtained, and data for Kheda are organised differently from those pertaining to Banaskantha. It is clear from Table 4.5, however, that in both the peripheral PHC areas the proportion of HGs that has been relieved of their duties is considerably higher than in the two core areas, leaving many villages without HG services. The issue of HG coverage is to some extent related to differences in the educational status of the village population. In peripheral areas, villages sometimes cannot propose even one candidate with the required educational background.

## 4.3 IMMUNISATION AND ORAL REHYDRATION IN THE PRIVATE SECTOR

In this study of the two technologies in the context of the EPI and CDD programmes, we have to be careful not to omit the activities in the private sector, as these can yield additional insight. In the case of immunisation, we have to distinguish between urban private doctors immunising infants for money and the immunisation campaigns which are sponsored and organised by organisations such as Rotary, Lion's, Tribhuvandas Foundation (TF) and Bansali Trust (BT). Here we shall focus on the work of TF, operating in Kheda district, and BT, operating in Banaskantha.

The TF was operating earlier in 175 villages in Kheda conducting a maternal and child health programme and training its Health Guides. It undertook immunisation in all these villages, but this was done rather haphazardly and the coverage was low. Poor coordination with the public health administration often led to over-immunisation as well as under-immunisation of infants. In 1985 the public health administration and the TF together decided that the latter would organise an intensive health programme in 60 out of the original 175 villages. These were selected because of their manageable size and peripheral location. Data on coverage would be sent to Nadiad and added to the target achievement of the district and of the PHC concerned. Vaccines would be provided by the government.

In Banaskantha, the BT is active in curative health care. It manages an eye hospital, a mobile clinic covering 10 villages, and carries out immunisation against polio. The last activity is carried out in camps organised during the period February to May by workers who receive training for ten days. Campaigns were held in 1980, 1983, and 1985. The programme is characterised by its *ad hoc* character and, compared with that of the TF, by a lack of adequate field structure. There is hardly any coordination with the public health administration. Only total coverage figures are sent to the authorities. In the field of oral rehydration the BT's activities are minimal. In a few villages under Varahi PHC they have appointed kindergarten teachers who are supposed to impart some health education to mothers and provide primary medical care to children.

Analysing the information we have collected on the work of non-governmental organisations, it seems that the NGO can find a niche in cases where the performance of government services is inadequate. It will stress this inadequacy as a justification for its own activities and for its growth. Its perception of government services will be geared towards 'black' spots and negative incidents. Cooperation with the public health authorities is based on exchange, for example, in Kheda vaccines and a legitimate niche in which to work against supplying coverage figures. This conclusion is very clear in the case of the TF, the NGO with the most organised programme working in the central district. In Banaskantha, the BT had a rather limited programme and only carved a very small niche, even though there seemed to be quite some scope in this peripheral district. Besides, the authorities themselves were aware of and, more or less, reconciled with the lacunae in their services. This seems to be the background for the relatively low degree of polarisation between the NGO and public services in Banaskantha. This does not imply, of course, that polarisation in Kheda was high in an absolute sense.

In the case of immunisation and oral rehydration, neither of the NGOs with considerable technical means (such as motor vehicles) and financial resources at their command, really offer a viable alternative implementation model. The demonstration value of their activities is, therefore, negligible.[7] They play, instead, the important role of covering the periphery, a role which is practicable as long as the government is unable to provide high level or full coverage through its own public health care organisation.

## 4.4 SCOPE AND PRIORITY OF WORK AT VARIOUS LEVELS

A survey of observations made at different administrative levels begins with the activities of the HGs.[8] We have already seen that only some of the villages have an HG. In immunisation, with its relatively high level of technical sophistication, HG services are virtually restricted to providing support when teamwork is being carried out, although there is no clear support pattern. We observed a few HGs guiding a team in their village, inviting women to come and bring their infants for immunisation. In one case, we even saw an HG holding a baby while it was being vaccinated, but this was clearly an exception. Quite often we were told by the MPWs that they never took an HG along. Though now and again a favourable remark was made, the predominant view among the MPWs was that the HGs could not be depended upon, as they were usually too preoccupied with other (non-health) work.

With regard to oral rehydration, the HGs are more independent. In the case of acute diarrhoea, they are one of the main sources of health care people rely upon. Their relative importance depends on the availability of alternative sources, such as a resident MPW, a well reputed doctor or traditional medical practitioner, or the proximity of a PHC. Visits to the HGs differed with regard to the nature of the services they provided. In Varahi PHC area the HGs had no ORS packets; whereas in the other three areas they did. Usually, the HGs would have a binding agent and some ayurvedic medicine. They were not very knowledgeable about the preparation of the home-made ORS mixture. One forgot to mention sugar, another forgot salt. However, the study's scope was far too limited to draw valid conclusions in this respect. On the other hand, the finding does support the general impression that ORT with a mixture of ingredients available in each kitchen does not get the emphasis it deserves in public health care delivery.

MPW activities are pivotal for immunisation. As we have already seen, they work in teams to achieve maximum coverage of the population. Not all of them, however, value teamwork in the same way. One male MPW said that he felt more secure when working with his female counterpart; several female MPWs stressed their preference for working on their own. From September onwards, all MPWs were to be on their own as far as immunisation is concerned—walking, travelling by bus and sometimes cycling, which only men do, to see their clients. In those areas where homesteads are scattered and villages far apart, it takes considerable effort to visit the people. The burden is even greater when the PHC is far away, as in the case of a female MPW who told us that she had to travel two hours to get to the PHC and two hours to get back by bus to get vaccines in the morning.

Dependence on vaccines and ice makes immunisation a vulnerable technology. Due to the sensitivity of the polio vaccine, it is a rule that these vaccinations are supervised during routine work. Generally, the MPWs wait till there are about 20 infants for polio immunisation before they ask the supervisor to come. One vaccine vial contains about 20 doses. On paper, polio vaccine is carried along during teamwork vaccination on a regular basis, but we heard of a case where the supervisor had to come down and the mothers had to be assembled on three extra days, as the vaccine was not available on the day vaccination had been scheduled.

The vulnerability of the technology is not the only factor which restricts the efficiency and manageability of the MPWs' work. When travelling in Gujarat we collected information on weekly or monthly working schedules of the MPWs at the PHCs, which we then used to locate them during working hours. Unfortunately, we hardly ever found them in the designated village on the designated day. The one time we did, they were engaged in work other than what they were supposed to be doing. We also had great difficulty in locating the residence of the MPWs. As it is not easy for outsiders, especially women, to rent a house in a village, they hardly ever lived in the villages they had been assigned to. The consequence of the rather haphazard choice of residence for MPWs is their uneven distribution over the rural population. Moreover, it complicates communication within the public health care services.

Studying the MPWs' immunisation work we were often struck by the intricacy of the pattern of factors influencing it among which *ad hoc* o‘The theory of those people is ɔrominent a role as structural factors. For example, the promotion of family planning is considered to be of overriding importance among the various activities MPWs are engaged in. Adopting the carrot and the stick principle, the priority for family planning from September onwards would obviously detract from attention necessary for routine activities like immunisation.

In 1985, immunisation was postponed due to a 73 days' strike organised by Gujarat public servants. When I visited the villages in August, no villagers could explain why, until then, no immunisation had taken place. No one in these villages seemed to have taken any interest in the matter which indicates that villagers are used to irregular visits from MPWs. A pattern emerges in which villagers are just recipients of what is offered and are not involved in shaping the system of public health care delivery.

Whenever I discussed the strike with MPWs and referred to the implication of putting infants at a risk by not immunising them, it was always pointed out that they were merely following the orders from Ahmedabad. They did not seem to feel any responsibility towards the children and parents in the villages where they worked. To some extent, this attitude seems to be representative of what I have called elsewhere the public health care culture prevailing in South Asia, characterised by the superseding of public responsibility by private interest.[9] But there is more to it. It also reflects the strong emphasis on the quantitative dimension of the MPWs' work. Their role is to achieve coverage targets and that is how they experience their work. MPWs visit the houses in the villages allotted to them usually only once a fortnight, which severely limits their potential to assist in cases of acute diarrhoea. Their main contribution to diarrhoea control is thus curing chronic diarrhoea. Further, as they were found to possess different combinations of drugs against diarrhoeal diseases, the MPWs interviewed differed in their potential contribution to curative care.

The PHC level, which comes next in the organisational hierarchy, is an essential link in the cold chain and in the vaccine delivery line. Moreover, training of lower level staff and supervision emanate from here. A PHC area is divided into sectors, each under a supervisor; and each sector in turn is divided into sections, each of which ought to be covered by a male and a female MPW. Due to vacancies, especially of female staff in remote places, the picture in reality is different. These vacancies often lead to neglect.

This was the case in a peripheral sector under Birpur PHC, which I visited on the day that immunisation had been planned. No team could be located, however, and the villagers I contacted were outspoken about the fact that no infant in their village had ever been immunised. When discussing these observations with the medical officer in charge of the PHC, he pointed out that this specific sector suffered in three ways: too few MPWs; a long-standing vacancy for a supervisor; and a recently appointed supervisor whom they had not seen after his first working day.

The PHC is the highest level actually involved in health care delivery. In comparison, the district health office is more of an exchange centre, from where aggregated coverage data are channelled to Ahmedabad and from where supplies are handed down. District level medical officers are, above all, implementors with the supreme task to attain the targets which the state administration has set. To accomplish this task, they use various incentives and are supported by supervisory staff. With regard to this staff, there are considerable differences between Kheda and Banaskantha districts. In the central district, there were two Public Health Nurses (PHN) who made their rounds among the PHCs, whereas in Banaskantha there were none. Two had been appointed from outside the district, but they refused to work, which they could afford to do because of some political influence. The medical officers at the district level have the public health training and epidemiological background necessary to investigate the health situation in their region and to formulate the most suitable policies. The present organisational structure, however, denies them such options.[10] Policies are formulated far away from the field, at the state level. The DHO's work in the fields of immunisation and oral rehydration is essentially that of a personnel manager and a high level bookkeeper.

In Ahmedabad the already aggregated coverage figures from the districts are collated and then compared to the targets set for the state. At this level, policies are formulated. Quite surprisingly, because of the vast distance between the state capital and the periphery, I came across a change in policy which had been based on interregional difference. A few years ago, a decision had been made regarding the requirements for the appointment of female MPWs. Initially, it had been decided that in remote areas local women would work in their own villages after being trained. But it proved to be impossible to find women with sufficient education in the periphery of Banaskantha. Hence, for that area alone, the rule was relaxed, allowing girls to go and work in a village other than where they lived

## 4.5 MANAGEMENT SYSTEMS AND MANAGEMENT INSTRUMENTS

We shall now look again at the implementation of the EPI and CDD programmes, but view both of them as a configuration of intertwined management systems. Each of these systems is geared towards reaching the objectives of the programmes and includes certain management instruments which are at the disposal of the public health care staff at various administrative levels. Basically, five management systems can be distinguished. The first two, namely, the cold chain maintenance system, and the vaccine supply system which are heavily dependent on technology, can seriously hamper the implementation of the programmes, even if the other systems run smoothly. Such problems often occur far away from the local implementation level, at the factory or the warehouse. Given the hierarchical structure of public health care administration, in such cases it is only the state level which may be able to get things going. However, the actual problem may be well beyond the influence of the state level public health administration, as is the case when vaccine production is interrupted by a strike or by a technical failure.

The third management system is the referral system. One of the pillars of a multi-layered health care system is the assumption that each level is specifically suited for treatment of cases with a certain degree of severity. Cases which cannot or should not be treated at a certain level must be referred to the next one up or down the line. For our study, the referral system is only of marginal importance. It does not play a role in the EPI programme, and the subject of cases of chronic diarrhoea or severe dehydration being referred to the most appropriate level, does not really concern us here.

The fourth management system is that of personnel management, comprising transfers and promotions. It is heavily intertwined with the fifth management system, the target achievement system, which includes the setting of targets, the instruments to boost achievement and the monitoring of what has been achieved. These two systems will be discussed in some detail.

### 4.5.1 The target achievement system

The main characteristic of the target achievement approach is its uniformity. Targets are calculated on the basis of demographic parameters and general principles developed at the international level. In no way are regional differences taken into account. On the contrary, strict adherence to a uniform strategy is supposed to generate pressure which will diminish the detrimental effects of, for instance, differences in quality of personnel and the vacancy ratio between areas. It further appears to be assumed that the approach will neutralise the implications of social differentiation and overcome cultural obstacles. How this is done and what negative side-effects may occur, are problems rather alien to this approach, as there is an absolute preoccupation with quantitative targets. Moreover, the system is extremely result oriented; responsibility, time and scope for research, experiments and analysis are reserved for the state and international levels of administration. The system requires an easily and clearly definable target population. Hence, it is most suited for preventive health care, like immunisation. In the case of curative care, where the population to be covered depends on the incidence of disease, calculation of targets and management of the achievement pattern are much more difficult.

The target achievement system is heavily dependent on incentives. In its most pronounced form this carrot and stick policy can be observed in the family planning programme. In this programme, at each level, prizes can be won, varying from trophies and medals to money, refrigerators and trips abroad. Prizes may be won not only by an individual MPW or medical officer at the PHC or district level, but also by the staff of a PHC collectively, or by the village panchayat.

During 1984–85 an effort was made in Gujarat to review the target achievement in family planning and in immunisation together when allotting prizes. Due to a shortage of vaccine, this procedure was not really successful, but continuation of this approach is intended. If it works out it will allow more room for positive incentives in immunisation. Presently, at all levels of implementation of the EPI, the emphasis is on punishment for non-achievement. Supervisors adopt a variety of methods to ensure target achievement—supervisory talk, show-cause notice, warning and stopping of increment. Moreover, there are the major punishments of demotion, transfer—officially this is not a punishment, but sometimes it is used as such in the target achievement system—and, ultimately, dismissal. Major punishments are not used in the target achievement system; minor punishments seem to be used sparingly with regard to the EPI.

Supervision is used more as a tool to exert control and check performance, rather than for motivation. Performance is mainly assessed by monitoring vaccinations: another management instrument. Given the bias towards target achievement, there is an in-built pressure towards 'positive monitoring', which means regarding a child as being immunised when that is not the case. Though it was not possible to scrutinise monitoring, we are of the opinion that this is not a big problem here. Actually, monitoring itself is checked in various ways, apart from regular supervision. As teams are involved in most immunisations, social control is an important check. Further, monitoring is not only meant to gauge achievement, but also to supply information on the vaccination status of children and pregnant women. All figures can be traced to names, and hence to that of a possible falsifier. Immunisation cards, which were maintained by mothers, had recently been introduced. These are another means to check monitoring.

Two management instruments remain. One is training, an aspect which did not receive much attention in the study, the other is evaluation. At the time I left Gujarat, a comprehensive evaluation of the EPI had just been completed which unfortunately, I did not succeed in acquiring. Evaluations had been previously undertaken in 1981 and 1982. Their character was, more or less, that of an extended coverage survey; recommendations were of a very general nature and did not include specific measures for future action.

### 4.5.2 The personnel management system

Medical officers at the district level are the primary advisors and decision-makers on personnel management. They give advice and take decisions on the filling of vacancies, transfers and promotions, though for MPWs there is little scope for promotion. On paper this gives them impressive power, but in reality the situation is different. Vacancies in remote places are not easy to fill, as it is exceedingly difficult to force people to move if they are able to muster political influence. Besides, such influence is also used when people try to move to what they consider a better posting. In fact, the personnel manager's hands are often tied, making the personnel management system an inadequate means for supporting the programmes to achieve their objectives.

There is also another dimension of the personnel picture. Several times I encountered health staff who had clearly found their niche in a peripheral place and did not want to move. Among the reasons mentioned were the fact that they belonged to that area, and one doctor even said that the peripheral nature of his PHC enabled him to carry on a private practice on the side. In such cases, the staff concerned would stay long in one place, building up strong rapport with the local population.[11]

## 4.6 THE ROLE OF THE ENVIRONMENT

The implementation of the programmes considered does not take place in a vacuum, even if this sometimes seems to be one of the assumptions of those who modelled them. This was most clearly expressed by a female MPW, who characterised the target achievement orientation as follows: 'The theory of those people is that more pressure leads to more cases, as if it is a vegetable market where buying is free.' Immunisation and oral rehydration of children take place in a cultural, social and geographical context, which influences the results considerably. The relation between the programmes and their context is, first, one of restriction. The environment imposes, for instance, limits on the 'reachability' of children. Second, the environment shapes the programmes. This may happen because of the permeation of the programmes by social principles or cultural orientations. A good example of permeation is the strike, where a reaction against a government policy of positive discrimination in a caste based society stopped timely implementation of the programmes from within. The environment also shapes the programmes through compliance. This occurs when the programmes are adapted to contextual factors, such as relevant ideas about the background and cure of a disease.

Third, the programmes shape the environment. In our case, health education may, for instance, change mothers' opinions about the value of preventive action. In this section I shall only discuss the first and second relationships. Lack of adequate data precludes pursuit of the influence of the programmes on society.

Let us, first, consider how the environment limits implementation. Most visible are the physical constraints. Lack of electricity hampers the maintenance of cold chains and bad roads add to the hardship of the MPWs. Remoteness is a concomitant of a lack of social infrastructure and makes it hard to fill vacancies. Hence, children in the periphery have lower chances of being immunised and receiving oral rehydration therapy when needed.

Labour migration restricts the possibilities of immunisation and oral rehydration therapy. Usually from September onwards, and sometimes earlier if the monsoon has failed, poor villagers from Banaskantha migrate to Saurashtra, and villagers from Panch Mahal migrate to Kheda. Because of this population movement a considerable number of children miss one or more vaccinations, depending on the timing of the migration and the coverage in the home village. There is, however, also another possibility. In order to raise their target achievement, Panchmahali children in temporary roadside settlements are sometimes immunised by Kheda public health staff. Kheda benefits in this manner, whereas Panch Mahal finds it slightly more difficult to achieve targets. Though the migrants are supposed to maintain vaccination cards, in reality a child often ends up with either too many or too few vaccinations. With regard to oral rehydration, temporary migration appears to pose fewer problems.

In addition to migration, there are other behaviour patterns which diminish the 'reachability' of children. First, a mother may refuse immunisation of her child and may even hide her child, not being convinced of the value of immunisation. Such a reaction may be either related to a previous experience or based on locally prevailing ideas on the background of disease. For instance, under Varahi PHC, there is a sizeable community of Ahir who rely on their own traditional health care providers and do not easily accept allopathic medicine. MPWs and their superiors recognise the greater difficulties of those working among the Ahirs, and accept low coverage among them. Eventually, with targets rising, the pressure to immunise Ahir children will increase and conflicts are inevitable.

Second, children may be hard to reach because women deliver in their parents' house and stay there for some time after the delivery. Though this custom is common, the duration of the stay differs between communities. The custom does not necessarily interfere with immunisation, provided that children absent during campaigns are vaccinated routinely, or immunised in their grandparents' village. But some fall between two stools, being still too young to be immunised when staying with their grandparents and going to a village where routine work suffers because of family planning activities.

Third, MPWs occasionally do not find children at home, as their mothers take them along to the fields. The impact of this behaviour pattern on coverage did not seem to be as large as I had expected. Some women said that they are informed of vaccination sessions in advance and try to stay at home or leave their child in the village. When I raised the issue of loss of income incurred by a woman staying at home, a Harijan mother said that a sick child would cost her more. In joint households it was quite common to find infants being looked after by older women, when their mothers were in the field. MPWs were aware of the problem and stressed that they tried to adjust the timing of the vaccinations.

Fourth, in some communities, such as the Thakors under Chandisar PHC, women usually remain in the house. It is extremely difficult for male outsiders to gain access to them and to their small children. Like the Ahirs, the Thakors are not found in all villages, but are concentrated in some villages where lower immunisation coverage is to be expected particularly when male MPWs are involved.

Fifth, public health care provided by government is only one of many sources of health care villagers use. The activities of NGOs and qualified medical practitioners have already been mentioned. There is also a large category of legally unqualified providers of modern Western drugs in the region. I met one of them in Varahi PHC, in a comparatively prosperous village. He had a degree in Ayurvedic medicine, but dispensed only allopathic drugs. In diarrhoea cases he always prescribed drugs; oral rehydration therapy was not recommended. There are many more sources: pharmacies in small towns, Ayurvedic practitioners, Unani practitioners, folk healers as among the Ahirs, and, in the case of a common ailment like diarrhoea, a large stock of home remedies.

Having sketched the elements of the environment which impose limits on the implementation of the EPI and CDD programmes, we can now review the ways in which the programmes have been influenced by environmental factors. As already mentioned, this may occur through a process of 'permeation', and I had expected that this would be true of the accessibility of services to socially underprivileged people, such as the Harijans. When I inquired about the quality of distribution among Harijans in all the villages visited, I was always assured that their part of the village was regularly visited by public health staff. The explanation can probably be found in the combination of a gradual erosion of the not too strong untouchability stigma in the region, with an emphasis on universal values in public health services, and with the pressure of the target achievement approach. I observed one case of possible neglect among a tribal (Bhil) community living on the outskirts of a large village under Alindra PHC, but it could not be ascertained what the matter was, as the grandparents were vague about the matter and the parents were away at work.

I observed two cases of permeation where the environment had led to a revelation of the programme rules. One rule requires MPWs to live in the village where they work, but this rule was relaxed when MPWs are not able to stay as they do not have a sufficient caste and kinship basis. Another rule to transfer MPWs was also relaxed as villagers occasionally want a female MPW who is due for transfer to stay on. Other changes included efforts to take into account the absence of mothers when planning immunisations.

## 4.7 CONCLUSION

At the beginning of the study, three problems were formulated for which findings will now be presented. The first problem refers to the ways in which the general objectives of the EPI and CDD programmes are translated into concrete action at the local level. To appreciate the findings with respect to this problem, it is crucial to know which options are available to the decision-makers. Both immunisation against six childhood diseases and diarrhoea control have to be accomplished on a continuous basis for technical medical reasons. A one-off campaign, as in the case of smallpox eradication, was not a feasible proposition.[12] Still, that leaves a major option. One possibility was to have mobile vaccination teams and, in the case of diarrhoea control, large campaigns to promote ORT. However, mobile teams do not fit in with the prevailing health care organisation and ideology in India, in which various kinds of services are integrated. The option of ORT campaigns, in combination with the routine approach, still seems feasible. This will remain so when, in the years to come, the emphasis will increasingly be on all four pillars of diarrhoea control.

In the case of the EPI, the general objectives have been narrowed down to explicit tasks for individual workers, which have been quantified along the line. The conversion of general objectives into quantified tasks is a major characteristic of the implementation process. In the case of the CDD, the emphasis has been on case management, a curative approach which leaves little scope for quantification. Here, the alertness to recognise cases and the transfer of knowledge to villagers are essential elements. I have the impression that the second element has not been well developed. One reason is that activities falling under the EPI and FP programmes tend to absorb the workers' attention more. In itself, this is a reflection of the lesser attention the CDD gets at all levels, because it is technically less spectacular and medically less prestigious and also because there is less political pressure from the (inter) national top of the hierarchy.

In the implementation of both programmes, the private sector plays a significant role in the towns, where there is a commercial demand for vaccinations and ORS packets, and in the rural periphery, where it fills up the gaps left by the public services. In the countryside the relation between the two sectors may be characterised as one of 'love and hate'. As yet, they need each other, but each success of the private sector demonstrates the weakness of the government services.

The implementation of services through the regular network mirrors the weakness present in the existing system. This can be seen most clearly in the relative neglect of people living in peripheral areas. On the other hand, the regular network has its strong points too. Implementation has, for instance, benefited from the fact that some necessary values, especially the target orientation and the emphasis on continuity and cyclical sequence of action, already prevailed in the network.

The second problem pertains to the management instruments which are used to attain a high level of achievement in the programmes. Five management systems which influence the results of the programmes were isolated. Two are geared towards technical aspects: the vaccine supply system and the cold chain maintenance system. The smooth operation of both these systems is essential, but breakdowns occur and these cannot be prevented, nor can their repair be speeded up much by the public health care system. The referral system, linking different levels of health care, was not studied, but there was an examination of the two remaining systems: the target achievement system and the personnel management system. It was noted that personnel management is instrumental in obtaining an equal coverage of services in all places. It was also made clear that the hands of the personnel managers are tied, as political influence proves to be a strong intervening factor in the case of transfer and promotion. The target achievement system is especially influential in the EPI programme. Its influence will probably become even more pronounced, due to the linking of the FP and the EPI target achievement at all levels, which will lead to further articulation of the 'carrot and the stick' management principle in the EPI.

In both the two technically oriented management systems, as in the personnel and target achievement management systems, monitoring and supervision are, besides 'the carrot and the stick', the major instruments used. Both instruments, with supervision essentially implying 'checking on performance', do fit in with the hierarchy and quantitative orientation, which are characteristic of the EPI. The CDD activities, with their emphasis on the transfer of knowledge, seem to have other management requirements. As can be observed from Table 4.1, the CDD technology is well suited for management by the people themselves. Such management on the demand side could also be a useful addition to the implementation of the EPI. If the villagers recognise, appreciate, and voice their right to immunisation, this would create pressure to pay more attention to peripheral areas. Additionally, such pressure might be used when lack of vaccines or breakdowns in the cold chain hamper the smooth operation of the programme.

The third problem is with regard to the relation between the programmes and their setting. It was mentioned earlier that the environment can limit the programmes but it can also change them, either by permeation, or by compliance. Though we found some examples of compliance, meaning that the programmes have to some extent been adjusted to fit the context better, there is more scope for this. The present low degree of compliance is in line with the present level of the EPI coverage and the relative neglect of the CDD activities. Things will become different when increasing pressure to attain full coverage for the EPI emanates from the international level. The same can be said about the level of diarrhoea control when more attention comes from above, which is bound to occur. For both programmes, further compliance seems to be the best future policy. For the EPI, this means that more effective ways must be found to cover the periphery and to increase the coverage for the third dose of a DPT immunisation course everywhere. For the control of acute diarrhoea, local knowledge can be better used. In both cases an effective use of evaluation, other studies and tools which are underused at present, is required. In addition there is a need to enlarge the room for manoeuvre at the district and, perhaps, PHC levels, so that the lessons of research can be put to better practice. Finally, there is a need to strengthen the HGs level, as these workers are continuously available in the villages. They have the rapport necessary to support the villagers in managing part of the CDD, and in developing a knowledgeable demand for immunisation.

## NOTES

1. The author would like to thank all those in the Directorate of Health, Medical Services and Medical Education in Gujarat, who provided their full support in carrying out the fieldwork; the IDPAD for its financial support; Professor Nirmala Murthy of IIM, Ahmedabad, who helped in getting the fieldwork in operation; and my colleagues Jarl Chabot, Jan Eshuis and Eric van Praag for their support in the epidemiological studies.
2. WHO is World Health Organisation; UNICEF is United Nations Children's Educational Fund; ICDDR(B) is the International Centre for Diarrhoeal Diseases Research in Bangladesh; CDC is the Centre for Disease Control, Atlanta, USA.
3. GOBI stands for Growth Monitoring, Oral Rehydration, Breast-feeding, Immunization. It is a selection of components from the broader, comprehensive Primary Health Care strategy formulated in Alma-Ata in 1978. At the moment, an international debate is going on about the merits and negative consequences of selective primary health care, which is mainly promoted because of its rapid impact and its cost-effectiveness. According to many,

TABLE 4.6

**DPT Coverage in Four Selected PHCs**

| *Year* | | | | *Varahi* | | | | | | | *Chandisar* | | | |
|---|---|---|---|---|---|---|---|---|---|---|---|---|---|---|
| | *Target* | *1st dose* | % | *2nd dose* | % | *3rd dose* | % | *Target* | *1st dose* | % | *2nd dose* | % | *3rd dose* | % |
| 82/83 | 2,150 | 1,750 | 81.4 | 1,631 | 75.9 | 1,441 | 67.0 | 3,840 | 3,172 | 82.6 | 2,809 | 73.2 | 2,548 | 66.4 |
| 83/84 | 2,140 | 2,240 | 104.7 | 1,985 | 92.8 | 1,172 | 54.8 | 3,835 | 3,287 | 85.7 | 3,001 | 78.3 | 2,786 | 72.6 |
| 84/85 | 2,570 | 1,958 | 76.9 | 1,576 | 61.3 | 998 | 38.8 | 4,590 | 4,390 | 95.6 | 3,710 | 80.8 | 3,414 | 74.4 |
| *Year* | | | | *Alindra* | | | | | | | *Birpur* | | | |
| | *Target* | *1st dose* | % | *2nd dose* | % | *3rd dose* | % | *Target* | *1st dose* | % | *2nd dose* | % | *3rd dose* | % |
| 82/83 | 4,897 | 5,012 | 102.3 | 4,389 | 89.6 | 3,690 | 75.4 | 4,729 | 5,501 | 116.3 | 4,554 | 96.3 | 3,076 | 84.1 |
| 83/84 | 4,950 | 3,138 | 69.5 | 3,007 | 60.7 | 3,063 | 61.9 | 4,710 | 3,294 | 69.9 | 2,588 | 54.9 | 2,170 | 46.1 |
| 84/85 | 5,923 | 3,844 | 64.9 | 3,525 | 59.5 | 3,378 | 57.0 | 5,635 | 3,503 | 62.2 | 3,103 | 55.1 | 2,593 | 46.0 |

**Source:** Monitoring data acquired at selected PHCs.

however, it will be self-defeating in the long run, as it is detrimental to the operation of the public health services.

4. For background reading, see Jobert [1985], and Maru [1985]. For the organisation of the public health services in Gujarat, see *Health Statistics of Gujarat* [1984].
5. For background reading on cold chain, see Dick [1985], which offers useful information on other aspects of immunisation as well.
6. See the *Action Plan for Oral Rehydration in Gujarat State* [1984] and World Health Organization [1981]. Though case management is most emphasised at present in Gujarat, chlorification of wells by HGs should also be mentioned.
7. See McCord [1976], about the requirements of a demonstration project in the health field.
8. General information on HGs is provided by Jobert [1985]; Maru [1985]; and NIHFW [1979].
9. Streefland and Varkevisser [1985, pp. 14–15]. See also Frank de Zwart's contribution to this volume for an interesting discussion of the attitudes of public health doctors.
10. This point is elaborated by Satia [1981].
11. Conversely, the local people also make more claims on the health worker or doctor over time. As a result, the village may try to keep the health care provider, or a doctor may be forced to move away as it implies a dwindling of his earnings. See Van der Veen [1981] and Frank de Zwart's contribution in this volume.
12. Smallpox eradication in India is described in Basu *et al.*[1979]. It is a pity that this well-documented volume devotes only two pages to contextual matters, under the heading 'Cultural and Religious Resistance', pp. 112–14.

# 5

# GOVERNMENT AND PRIVATE HEALTH CARE: TWO COMPETING INSTITUTIONS

**Klaas W. van der Veen**

## 5.1 INTRODUCTION

This chapter analyses the consequences for tuberculosis (TB) treatment of a strategic management decision in India to establish state organised health care services parallel to an expanding private sector after Independence.[1]

The funding and organisation of health services in India have, as in many other countries, been strongly influenced by their colonial past. This explains the type of attempts that have been made to provide free medical services.

In Western countries there has been a gradual transition from private health insurance to universal, socialised services, usually financed by considerable contributions from employers and employees for health insurance schemes.

The idea of socialising health services in India has been implicit in government health planning since the recommendation of the so-called Bhore Committee in 1946. Obviously, the economy does not yet allow the free provision of medical services to the population at large. The result is state organised health services working parallel to a large private health care sector. The effects of this situation will be illustrated by examining the National TB Programme in a district in Gujarat. The analysis shows that the role of state and private sector in TB care are certainly not complementary, but, instead, highly competitive. This situation, which is typical of the organisation and management of health care programmes in India, is demonstrated here by the detailed presentation of one example. The example clarifies how the opposition between public and private medical sectors adversely affects the health of TB patients and impedes the success of the TB Programme in general. Before the detailed case is presented general background information on the financial structure of health care is necessary.

## 5.2 STATE AND PRIVATE HEALTH CARE ORGANISATIONS

### 5.2.1 General historical background

Over a century ago, in 1883, Germany led the way for the 'socialisation of medicine' by instituting compulsory sickness insurance for all workers. An existing organisation of separate, autonomous, mainly occupationally based health insurance funds was brought under the joint control of employers and employees. Fixing the terms of service for doctors was determined by the funds and not the government. In other European countries similar systems evolved. In Britain, the first comprehensive legislation was passed in 1911. France followed in 1928 and most other European countries had some sort of arrangement before the Second World War. In most of these schemes, doctors remained private entrepreneurs who were not, or only partially, in the service of the central or local government authorities.

In Europe, compulsory health insurance developed at a time when there was no evident shortage of doctors. The only services available free to the population in general were provided for the poor, either by way of charity or under the Poor Law, as in Britain since 1834. With very few exceptions, these services were of questionable quality, limited, and they stigmatised the user (Abel-Smith [1985], p. 961). People in regular employment were, therefore, prepared to pay insurance contributions to obtain the right to better services, rather than to pay their bills for private services as they needed them. This created a broad base for insurance systems in which both employers and employees took a share in the costs for health care. Compulsory sickness insurance is a first step towards the socialisation of medicine as the provision of medicine is considered a public service which should be made available to all people, irrespective of their economic condition, caste, creed or social status.

After the 1917 revolution, Russia was the first country to give its citizens a constitutional right to all health services and to curtail the private practice of doctors drastically.

In Britain, the National Health Services Act of 1946 also realised a full-scale socialisation of health services. It was the outcome of a gradual transition and not, as in Russia, part of a revolutionary change of the existing social order. In Britain, it was an extension of the original private and local sickness insurance funds and the involvement of local authorities on a nationwide scale. The central government also decided the terms of service for doctors.

In a few other Western countries similar national health services were realised. Compulsory health insurance was in one way or another integrated in the overall government legislation in these countries. Only after the Second World War did the notion that the government should assume responsibility for proper health care for the people become accepted.

### 5.2.2 State and private health care in India

With the British example as their frame of reference, it is not surprising that the architects of the Constitution of India wrote that 'the state shall regard the raising of the level of nutrition and the standard of living of its people and the improvement of public health as among its primary duties.'

A second political commitment, which turned out to be even more deeply held and overriding, was to bring about the desired changes in health services without making basic changes in the then existing social order and government machinery. This implied that the state health facilities were to run parallel to the clinics of private practitioners and voluntary organisations.[2]

In India, no tradition of private health insurance existed before 1947. The first health insurance schemes, based on the principle of contributions by employers and employees, were introduced shortly after Independence. The Employees State Insurance Scheme (ESI) for workers in the big industrial enterprises from 1948, was the first such scheme. This was followed by the Central Government Health Scheme (CGHS) for government servants in 1954.

The personnel of the Defence Services and Railway employees have similar health insurance provisions. These schemes involve large groups of workers in the formal and government employment sectors. However, the number of beneficiaries totalling 26 million people is less than 4 per cent of the total population (Park and Park [1983], pp. 595–96).

In India, as in most developing nations, the attempt to provide free medical services for the whole population has prevented the development of informal systems of voluntary health insurance that later could be integrated in the government services. It puts a tremendous financial burden on the government!

The Indian population has to choose, for all practical purposes, between private payment for medical services provided by practitioners who operate as entrepreneurs, or to rely on the services of charitable institutions or the government health care organisation.

There has been an enormous expansion of state organised and state financed dispensaries and hospitals. There are now about 26,800 dispensaries, 4,000 hospitals and 8,000 rural Primary Health Centres, compared with a few hundred widely scattered establishments in 1947. The number of medical colleges grew from 29 in 1947 to 106 in 1980, while the annual admission increased from 2,500 in 1950–51 to 13,000 in 1980. A minority of the graduates of these institutions, 25–30 per cent, seek government employment. The others settle as private practitioners and consultants in India. Some go abroad to the Middle East and until recently to Britain and the USA. Thus the expansion of the state sector is surpassed by that of the private sector.

The state organised facilities are mainly concentrated in cities and towns where only 25 per cent of the population lives; there is a shortage of funds, medicines, and of sufficient and well motivated personnel, rendering the services of the state sector a second choice for the people.

Since the recommendation of the Bhore Committee in 1946, the 'eventual creation of a national health service' is constantly discussed, but the socio-economic constraints are too high. This is expressed in the most recent recommendations for an alternative strategy in health care, 'the time is not ripe for such a measure . . . One cannot abolish the private sector in all important walks of life' (Health [1980], p. 143). The policy so far, and the proposals of the study group of the Indian Council of Social Science Research and the Indian Council of Medical Research, have therefore been 'to provide free health care for the poor, and under-privileged sectors of the population.' It was hoped that 'through the improvement of services there will be less and less need for a citizen to go to private practitioners and consequently . . . the domain of private practice will be considerably reduced' (Health [1980], p. 143). It is taken for granted that the state and the private sector will complement each other. This is sound reasoning for any one who has seen the opportunities for 'exploitation of illness' by private practitioners in a social setting without a full-fledged social security or insurance system.

There are vast individual differences in the extent to which doctors let mercy prevail over personal gain. Of course, private practitioners are also private entrepreneurs, and they cannot be blamed for first considering their patients' ability to pay. After all, the availability of government health care services enables patients, theoretically, to obtain free treatment. However, the crucial problem is that patients do not want to use these services.[3] This problem and its consequences will be discussed in detail after a short introduction to the development of the National TB Programme in India.

## 5.3 THE NATIONAL TUBERCULOSIS PROGRAMME

### 5.3.1 Background of the National Tuberculosis Programme

Between 1955 and 1958 the Indian Council of Medical Research conducted a National Sample Survey to study tuberculosis in India. This study revealed, rather unexpectedly, that tuberculosis was as rampant in the villages as in Indian cities and towns. It also appeared that, among the communicable diseases, TB accounted for the highest rate of morbidity and mortality.

In order to improve the situation the government established a National Tuberculosis Institute in Bangalore in 1959, and charged it with the task of developing a nationwide programme to combat the disease. This programme was merged in 1962 with the existing tuberculosis health services which were till then the result of the work of NGOs, though the government had a stake in the services too. The main NGO active in the TB health services was the Tuberculosis Association of India (TAI).

In 1947, there were 80 clinics and 6,000 beds available for the treatment of TB patients. At present, there are 600 clinics in the country of which nearly 360 are district centres organising district TB Programmes. There are still some 15 districts where no special facilities are available. In the remaining 40 districts not yet covered by a regular district TB Programme, it is, however, possible to organise TB control with a little effort (Nagpaul [1981], p. 572; Barua [1977], p. 36). The main objective of a district TB Programme is to organise case finding and treatment through the existing peripheral health institutions of a district.

Studies carried out in the fifties in India showed that the results of therapy at home with anti-TB drugs that had become available after 1950[4] were as good as the results of sanatorium treatment and that bed rest did not play a major role in the treatment of TB. In 1946, according to the Bhore Committee Report, the expectation was that some 500,000 beds would be needed for proper control of TB. But home treatment reduced the need for hospitals, sanatoria and beds. At present, there are nearly 8,000 beds available for the treatment of patients who are seriously ill with haemoptysis and spontaneous pneumothorax, or who do not respond to the standard anti-TB drugs.

### 5.3.2 Tuberculosis Association of India

The Tuberculosis Association of India (TAI) was founded in 1939 to coordinate various voluntary activities. In the first two decades of its existence it established clinics and sanatoria, organised programmes of health education, conferences and training. It also played an important role in research promotion. At present, there are 23 State Tuberculosis Associations affiliated with the TAI and since 1950 money has been raised through a large-scale campaign.

Since the introduction of the National Tuberculosis Plan (NTP), the role of the TAI has been seen as supplementary to that of the government. 'Associations should work to streamline the working of the (NTP) programme. They can assist government agencies in various fields where such assistance is called for' (Cariappa [1981], p. 594).

Since 1948 the TAI, through its Technical Committee, has played a very important role as advisor to the government. It has coordinated scientific research, launched the *Indian Journal of*

*Tuberculosis* in 1953 and organised all-India conferences every year. The publication of a *Textbook on Tuberculosis* in 1970 (second edition in 1981) was the result of the TAI's continuous efforts. There can be no doubt about the high level of contribution of the TAI to tuberculosis care in India. Unfortunately, the impact of the work of the State Tuberculosis Associations in the villages is very small. The Associations do not mobilise the population on a massive scale; they operate as charities.

In Gujarat, there are 11 NGOs operating non-profit TB hospitals where, contrary to the advice of the government, the NTP and the TAI, patients are hospitalised for periods of one to three months on an average. Since the official approval of home treatment as the best and the most economical approach to the tuberculosis problem, the number of beds in such institutions has doubled in Gujarat! This is contrary to the national trend.

The 11 voluntary hospitals in Gujarat, which are all concentrated in the northern and central regions, are not officially affiliated with the TAI. There are clear signs of competition between these NGOs and the government, and between the NGOs and the state branch of the TAI, which is clearly identified with the government by the people.

### 5.3.3 General results of the NTP

Despite all its efforts, the impact of the NTP is minimal. Today tuberculosis is a curable disease: estimates about the efficiency of modern chemotherapeutics are as high as 95 per cent.[5] But the NTP has remained 'a rather restricted service'; the 'incidence and prevalence of the disease has not shown substantial reduction, although the severity of the cases has been reduced due to effective drugs', and it 'still ranks high as the single communicable disease causing largest morbidity and mortality' (Health [1980], p. 99). The district Programme is still functioning at 30 per cent of its potential in case finding efficiency (Bailey [1982]). On a national scale, only 15 per cent of patients with possible TB symptoms go directly to government health centres on their own initiative (Pamra [1980]). The NTP, designed as a 'felt need' programme, offers treatment free of charge, but fails to reach the masses.

Economic conditions are the main reasons why India has failed to cope with tuberculosis effectively. The disease thrives where there is poverty and people suffer from poor nutrition and live in cramped, unhygienic housing conditions.

Social and cultural factors are other reasons why TB is still one of India's major diseases. The complex nature of these social and cultural characteristics is discussed in the next section. The results of a study are discussed in order to elucidate the impact of these variables on the treatment of TB.

## 5.4 PROBLEMS OF TB TREATMENT IN A DISTRICT IN GUJARAT

### 5.4.1 Material and methods

The study was carried out in Valsad, the southernmost district on the Gujarat coast, where the literacy rate of the population is in the top 15 per cent of the country. The coastal plain of the district is densely populated and has a well developed transport and communication network. The district is predominantly agricultural, though there is substantial industrial development around the towns. Access to medical facilities in the district area is relatively easy compared with other parts of Gujarat and India.

The study was carried out in two phases. During the first stage, the author, who had been researching in South Gujarat for almost two decades, surveyed tuberculosis patients living in one village with a population of about 5,600.[6] An intensive, qualitative analysis of the history of illness and treatment of nearly 40 patients was carried out for two reasons: (*a*) to develop a theoretical framework for a larger quantitative study, and (*b*) to design a questionnaire for this larger study. The focus was to identify patterns of treatment in relation to the socio-economic position of patients and the choice between private practitioners and public health facilities.

The beliefs of the patients about the best way to become cured was a main theme during the first stage of research.

During the second stage of the study in 1981, Kashyap Mankodi, from the Centre for Social Studies, and Professor N.R. Mehta, at the Department for Social and Preventive Medicine from the Government Medical College in Surat, selected 12 villages in Valsad district from which 959 patients had been registered with the District TB Centre (DTC) since its inception 12 years ago.

In these villages a representative sample of 569 families was interviewed. The data thus obtained provided a socio-economic profile of the villages and of the health care characteristics of the inhabitants.

Another source of information was treatment data of all TB patients from the 12 villages who had been registered at the DTC. The researchers collected patients' histories and tried to trace the patients and interview them. The results of this procedure are presented in Table 5.1.

TABLE 5.1

**Information on 959 Patients from Twelve Villages Registered at the DTC**

| | *Total* | *Percentage* |
|---|---|---|
| Interviewed | 402 | 41 |
| Migrated | 165 | 17 |
| Dead | 232 | 24 |
| Untraceable | 153 | 16 |
| Refused Interview | 7 | 2 |
| Total Registered | 959 | 100 |

The interview schedule elicited routine socio-economic information from the patients, and their detailed history of illness, the measures taken to cure it, and treatment by the DTC and private practitioners. The beliefs and opinions of the patients about TB and their health status were also examined during the interviews.

### 5.4.2 Limitations of the sample and additional information

The patients interviewed did not represent all TB patients who had registered at the DTC, representing only non-migratory residents. They probably have a greater resistance to TB or a better acceptance of treatment because they had survived the disease.

Another drawback of the sample was that it included only those patients who had at some time registered with the DTC. For practical reasons, this was the only group that could be traced comparatively easily. Even though addresses were noted from the DTC registration cards, it took a long time before the patients or their relatives could be contacted; 33 per cent of the registered patients could not be traced at all.

Furthermore, as most private practitioners do not register their patients it was virtually impossible to obtain an acceptable sample of patients who had sought treatment from private practitioners only.[7]

However, the sample was large enough to analyse quantitatively the utilisation of medical facilities and the patterns of treatment by different socio-economic groups.

## 5.5 RESULTS AND DISCUSSION

The findings of the study will be discussed under three major headings. The outcome of the anthropological pilot study; the performance of the DTC; and the characteristics of the DTC patients.

### 5.5.1 The outcome of the anthropological pilot study

Conversations with patients, their relatives and doctors in one village revealed the widespread belief that it is much better to consult private practitioners than to visit the DTC or a Primary Health Centre for TB treatment. Even patients who had been treated from the very beginning at the DTC and who had been effectively cured, said that they would have preferred a private practitioner if they had the money.

Repeatedly, we heard about patients who had received several short courses of treatment from private practitioners, usually over a period of one to three months and which was repeated over several years. The short course of treatment seemed to yield good results in the initial stage, but later on the illness recurred and short therapies were no longer effective. Such patients became chronic fibrotic cases not amenable to any drug therapy.

With modern drugs, apparently miraculous relief from symptoms is possible within weeks after starting the treatment. This happens particularly in the first overt stages of tuberculosis. Hence, very few private practitioners inform their patients at the outset that they will have to continue treatment for a long time. As one of them said: 'I would be suspected by my patients of greediness if I were to tell them at their first visit that they have to take treatment continuously for a year or more.' Furthermore, 'success' stories of people who felt much better within a couple of weeks are very common. All doctors know that their patients will go elsewhere when they do not suggest 'within a few weeks you will feel much better'. Even if private practitioners do advise patients to continue treatment, the patients will discontinue the treatment because of lack of money. The doctor can easily blame the patient who does not care for good advice and good treatment. This situation has led to a popular stereotype of 'good doctors' who can cure patients within a couple of months. It also leads to a great number of patients who became immune to drug therapy.

The histories of several patients revealed that private practitioners who had treated these patients had advised them to visit the DTC. It is, after all, negative propaganda if a patient dies when he is still under treatment.

This pattern of on-off treatment and the consequent recurrence and the chronic nature of the disease thus, oddly enough, reinforces popular beliefs about the efficiency of private practitioners. The referral of advanced and virtually untreatable cases to the DTC, where they often die, perpetuates the poor image of the public health system. The quantitative data, however, of the survey reveal a different picture.

### 5.5.2 Performance of the DTC

According to the estimates based on the National Sample of Tuberculosis carried out between 1955 and 1958, there was an estimated incidence of 4 per 1,000. The average percentage of cases detected by the DTC between 1970–80 was 38.3. Pamra [1980] reports an even smaller all-India figure of 15 per cent. Both figures indicate that the majority of TB patients do not even attend the referral system of the NTP.

To evaluate the DTC's performance, information available for 959 patients was examined: 42 per cent of the patients had been treated for a period of less than three months and only 25 per cent had been treated for one year or more. The treatment of only 11 per cent had been terminated by the DTC because the patients had been cured.

It is questionable who is to be blamed for the very poor performance of the DTC and the alarming number of 'early defaulters'. Is it the bad services of the DTC? or is it the idea that patients can be cured in a few months by private practitioners? This myth has never been refuted by private practitioners who are often the first to be consulted by the patients.[8]

### 5.5.3 Who goes to the DTC?

The (ex-)patients of the DTC numbering 402 who had been interviewed showed the following characteristics.[9] They were poorer than the average population; 58 per cent lived in huts compared with 46 per cent of the people in the 12 selected villages; 82 per cent had no land compared with 71 per cent of the people in the 12 villages.

It was very striking that illiterates and landless agricultural labourers, two overlapping categories which represented the poorest section in the villages, were underrepresented in the sample: 16 per cent of the DTC patients were agricultural labourers and 9 per cent were illiterates; in the survey of the total population these percentages were 21 and 14. They are apparently too poor and too dependent to be ill! Furthermore, they obviously have limited access to the state organised health institutions which are especially intended for them. The awareness that they can utilise the services provided by these institutions is not actively developed by their employers nor by the personnel in the peripheral DTCs. The majority of the agricultural labourers who do visit the DTC, report their illness at an already advanced stage and go there directly. They had not initially consulted private practitioners of allopathic medicine.

The physical condition of agricultural labourers reporting at the DTC is poor: 36 per cent of these labourers suffered from extensive bilateral lesions in the lungs, compared with 20 per cent of the 402 persons interviewed; 69 per cent of these interviewed labourers were unable to work any more, as against 52 per cent of the 402 DTC patients. The conclusion by Banerji [1981] that the suffering caused by TB is overwhelmed by the suffering caused by other problems such as hunger and abject poverty, appears to be supported and is applicable to agricultural labourers in particular.

The fact that a large group of patients are advanced cases is indicated by 52 per cent of the interviewed DTC patients who were unable to work any more when they reported at the DTC. These patients often had a medical history of on-and-off treatment by private practitioners.

In the sample of the 12 villages, only 27 per cent of the TB patients had gone directly to the DTC, 73 per cent of them had first consulted private doctors. Twenty-eight per cent of those who had first consulted private doctors had even irregularly visited two or more other private practitioners over a long period of time before registering at the DTC.

In order to find out whether patients had been examined or treated for TB, they were asked (*a*) if they had been told that they had tuberculosis; (*b*) if they had been advised to take prolonged treatment; and (*c*) whether they had screening or sputum examinations.

A comparison of the diagnosis and treatment of patients who first visited private practitioners prior to visiting the DTC, and that of the same patients at the DTC is most revealing, as is shown in Table 5.2

TABLE 5.2

**Diagnostic Action Taken and Advice Given to Patients at their First Visit to Private Practitioners Compared with the Same at their First Visit to the DTC**

| *Diagnosis and Advice* | *Private Doctors* | *DTC* |
|---|---|---|
| | *(Percentage)* | |
| X-ray taken | 22 | 93 |
| Sputum examination | 8 | 79 |
| Advised prolonged treatment | 15 | 74 |
| Patients informed that they had TB | 20 | 73 |

The lack of a proper control group of patients who had only been treated by private practitioners blurs the overall picture. Still, the data clarify that the effectiveness of diagnostic actions taken by the DTC is far superior to those taken by private practitioners. Furthermore, and even more important, it appears that at the DTC many more patients are informed about the necessity of prolonged treatment.

Information about the length of the treatment is crucial for the success of any TB programme. In most cases, private doctors are the first to advise the majority of TB patients. They have a key role in the success of the treatment, which they obviously do not fulfil properly.

The private sector interferes in a most unfortunate manner with the state organised TB programme. Private practitioners effectively complicate the problem of patients who do not continue drug treatment because they do not undertake any action to contact them. This is probably due to the lack of motivation among doctors to do so, which could also explain the virtual absence of records of their patients. By not emphasising from the very beginning the importance of prolonged and regular treatment, doctors are able to capitalise on the poor stereotype of government institutions and the lore of the 'good doctor'.

## 5.6 PRIVATE PRACTITIONERS, THE TUBERCULOSIS ASSOCIATION OF INDIA AND THE NTP

The malignant role of private doctors is one of the consequences of the parallel organisation of the private and state health care sectors. The government pretends that the two are complementary, each with its own clientele and tasks. The official documents on the coordination between government services and private practitioners, for example, in the most authoritative source concerning tuberculosis, the *Textbook on Tuberculosis*, illustrate this belief:

> There are always a large number of private practitioners in big cities. No matter what facilities are made available in any tuberculosis clinic for diagnosis and treatment in metropolitan cities, some patients will always find it more convenient to take their treatment from a private practitioner. Quick disposal and a personal relationship with the physician (as against an impersonal contact with different visits to the clinic) are some of the motives for preferring to go to private practitioners even if it is more costly. Steps must, therefore, be taken to bring in private practitioners in the programme of tuberculosis control rather than their competing with the clinics (Cariappa [1981], p. 590).

The research results clearly indicate that there are not 'some', but many patients who consult private doctors. The vast majority of TB patients first consult private practitioners, who do not give them proper instructions and their chances for full recovery are hindered. There is every reason to involve private doctors in the national health care programmes.

The Tuberculosis Association of India, which is the biggest voluntary association in India, can play an important role in guiding this integration. It is promising to observe how members of the Association envisage this integration of private doctors in the national campaign to fight tuberculosis. A lecture given by the Association for general practitioners in 1985 contains the following passage:

> For full involvement and support of private practitioners in the National Tuberculosis Programme, some facilities have to be given to them. Nearly 80 per cent of the total patients in the country live in the rural areas. Most of them are poor (since poor people are liable to suffer more from tuberculosis) and cannot fend for themselves. Nor they may be able to afford the time and money to attend the nearest Primary Health Centre, which may, sometimes, be as far off as 20 km. If private practitioners could also refer their patients for free sputum examination (and X-ray examination, if available) to the government health centres, it will help the rural patients tremendously and the diagnostic facilities will be immediately expanded. One may further suggest that the state may also make anti-tuberculosis drugs available to those private practitioners who are willing and are agreeable to maintain proper accounts for free distribution to their patients who cannot afford to buy the drugs. In other words, private practitioners' clinics should also be recognised as authorised sub-centres of the District Tuberculosis Centre for purposes of case-finding, drug distribution, etc. These steps will help to increase the facilities substantially and also help those patients who may find it convenient to consult private practitioners (Pamra [no date], pp. 33, 34).

Though tuberculosis is a specific case, similar problems will be prevalent in other official health care programmes.

## 5.7 CONCLUSION

The idea that TB treatment by private practitioners is superior to that provided by government institutions is not correct and this should be made clear. Medical students must be taught during their training that it is necessary to instruct TB patients about regular and prolonged treatment in order to be cured.

In turn, the government must make a conscious effort to involve private doctors in the National Tuberculosis Programme by offering them attractive remuneration. Effective strategic management calls for the open acknowledgement of the interface of private and state activities, an analysis and elucidation of this relationship and the taking of decisions in the light of such an analysis.

## NOTES

1. This chapter is based on fieldwork initially carried out with a grant from the Directorate General of Development Cooperation of The Netherlands and with later support from the IDPAD programme. Many thanks are due to Kashyap Mankodi of the Centre for Social Studies, Surat and Professor N.R. Mehta of the Medical College, Surat. Very useful comments were made on the first draft by the participants of a conference in Jaipur, March 1986.
2. There are in India different medical systems, such as folk medicine, Ayurvedic, Unani and homeopathic health care. The number of practitioners of these different approaches far exceeds the number of allopathic practitioners. Almost all the traditional medical practitioners are private doctors. In this chapter only allopathic health care facilities will be discussed.
3. Many studies have confirmed that the utilisation of government health services and Primary Health Centres in particular has been restricted to those living within a range of about 5 km of the government health institutions. Furthermore, the centres use only 20 per cent of their capacity (Rangamman *et al.* [1968] and Kothari *et al.* [1982]).
4. Streptomycin was the first really effective anti-TB drug which became available for general use in 1949. Since the introduction of streptomycin, several synthetic anti-microbials have been discovered. Most of them require a long period of regular use (Rao [1981], p. 8).
5. See Radakrishna [1983] for a discussion of the efficiency estimates in general and in India in particular.
6. See for a detailed account Van der Veen [1982].
7. Since the study also sought to examine the interaction between the private and public sectors in case detection and case holding, comparable information about patients who had never registered at the DTC was needed. Therefore, in one of the selected villages a team of 11 interns from the Primary Health Training Centre surveyed 900 households whose members had been medically examined. Fifteeen patients were traced who had been treated for TB by private practitioners only. The data gathered confirm the general trends discussed in this paper.
8. During a series of lectures on this topic at Medical Colleges in Gujarat I found that even professors at Government Medical Colleges were quite convinced that it was better to refer patients to private practitioners.
9. The results of the diagnostic activities at the DTC showed for the 959 registered patients, respectively, 24 per cent sputum and radiologically positive, 64 per cent only radiologically positive, and 12 per cent extra-pulmonar tuberculosis. The results of the 402 interviewed patients were: 23 per cent sputum and radiologically positive, 64 per cent only radiologically positive, and 13 per cent extra-pulmonar tuberculosis. For a more general discussion of the data see Kashyap Mankodi and Klaas W. van der Veen [1985], and Van der Veen [1981].

# 6

# CENTRALISED AND DECENTRALISED RURAL DRINKING WATER SUPPLY SYSTEMS AND THE PUBLIC–PRIVATE NEXUS

**Hein Streefkerk**

## 6.1 THE PROBLEM

In Gujarat almost 60 per cent of the 18,275 villages face acute water scarcity, either because the sources are dry or too distant or the water is saline, or polluted (Mehta [1984]). These villages are officially categorised as 'problem villages' or 'no-source' villages.

In Gujarat, two administrative institutions are responsible for planning, financing, constructing, operating and maintaining the rural drinking water supply: the panchayat administration and the state government. The former is a three-tier system of local government with extensive responsibilities vested in three locally elected bodies: the village panchayat, the taluka panchayat, and the district panchayat. The last one constitutes the highest echelon in the panchayat structure. The taluka panchayat is composed of elected representatives from the village panchayats; and the district council consists of chosen representatives from the taluka panchayats. The responsibilities of the panchayat administration include maintenance of sanitation, health and public works, expansion of educational facilities, improvement of agriculture and running the village administration. The main financial sources of the panchayat administration are land revenue and government grants transferred by the state government for the implementation of programmes in the spheres of agriculture, education, sanitation, and health. Other resources are taxes and fees, and donations and gifts that the village panchayats receive from the public.

The state government administration consists of the state legislative assembly, the cabinet, and ministerial departments such as the Health and Family Welfare Department, Industries Department and Judicial Department. Apart from these departments, there is a variety of more or less autonomous 'boards' or 'corporations', such as the Gujarat Water Supply and Sewage Board, Gujarat Industrial Development Corporation, or Gujarat Agro-Industries Corporation. They all implement policies in their respective fields. They operate and maintain important public facilities like the State Transport Corporation and the Gujarat Electricity Board. The head offices of these departments, boards, and corporations are located in Gandhinagar, which is situated about 30 km from Ahmedabad. The Gujarat Water Supply and Sewage Board (henceforth the Board), the state organisation which is responsible for the provision of drinking water, was founded in 1980–81. Before 1980, the Board's current organisational apparatus was part of the Gujarat Health and Family Welfare Department. One of the tasks of the Board is to support municipalities in the construction and maintenance of drinking water and sewage systems. Another important task is to provide 'no-source' villages with drinking water. This is done in those areas where the villages, or rather the panchayats, are unable to provide water for technical or financial reasons. It runs regional supply schemes whereby many villages are connected to one or several common sources. It also has 'individual' village schemes, where village specific facilities are provided.

In this chapter the introduction of drinking water in peripheral and central areas[1] will be compared. The aim of the comparison is to formulate hypotheses on management problems and their solutions. In other words, do problems of construction, operation and maintenance of recently introduced drinking water schemes differ in peripheral and central areas? If so, what are the factors that account for these differences? Finally, are these problems such that they could be remedied by improved management procedures?

In practice, a comparison between central and peripheral areas resulted in a comparison between more and less centralised rural water supply systems. In the central areas, the village panchayats are responsible for the supply of drinking water. In peripheral regions, the Board is responsible, either by organising regional supply schemes or by providing village specific facilities. This comparison of different types of water supply systems, instead of central and peripheral areas, deviates somewhat from the original research design. However, this methodological complication is not very serious because in the central areas one often finds decentralised schemes. In peripheral regions centralised facilities dominate. The Board provides those areas and villages with drinking water where the existing sources are inadequate or the panchayat is unable to improve them because of technical or financial reasons. These remote areas are generally economically poor, suffer from poor physical conditions, insufficient infrastructural facilities and are not manned by the best administrative personnel.

Geo-hydrological conditions and the consequent technical and financial requirements largely determine whether water supply systems are more centralised or not. If villages depend on distant sources, the waterworks will be more centralised than those in areas where each village can rely on its own local sources.

The extent of centralisation or decentralisation also depends on whether decisions and actions regarding planning, finance, operation and maintenance are taken at the national or state capital, at the district headquarters, or at the taluka (subdistrict) and village levels.

Drinking water systems can be viewed on a continuum that ranges from more to less centralised. At one end are the rural drinking water supply schemes run by the Board for which the major financial decisions are taken in New Delhi. How they are planned, constructed, operated and maintained is decided at Gandhinagar. Their actual implementation is delegated to lower levels of authority. The Deputy Executive Engineers in the branch offices of the Board, who constitute the lowest level of authority, are responsible for the implementation at the taluka level which includes many villages.

At the other end of the continuum are the village specific supplies. They are planned, operated and maintained by the village panchayat with the assistance of the district panchayat in matters of finance and construction. In between these two extremes are the village specific waterworks which are planned, financed and constructed by the Board, but operated and maintained by the village panchayat. In the study on which this chapter is based, Banaskantha district is considered the periphery and Kheda district the centre. The former is located in North Gujarat, bordering Rajasthan in the North and Pakistan in the West. Its western part, where Santalpur and Radhanpur talukas are located, is the poorest area of the district. Drinking water is supplied mainly by regional schemes which have been recently introduced. Palanpur taluka and Palanpur town, the district headquarters, are located in the eastern part of the district. In Palanpur taluka there are village specific waterworks run by the village panchayat and financed and constructed by the Board.

Kheda district is located in Central Gujarat with Nadiad taluka as its centre. Balasinor taluka, in the northeastern part, is the periphery of the district. In Nadiad taluka, the majority of the village waterworks are run by the village panchayats. The Board has financed and constructed drinking water facilities only in a few villages. In Balasinor taluka, drinking water systems are run by the village panchayats, though several of these systems have been provided by the Board.

The implementation of a drinking water scheme in Banaskantha district will be discussed first. Then, the village waterworks in Kheda will be analysed, and finally the main conclusions will be presented.

Data were collected during May and June 1984 and during August and September 1985. I spent three weeks in Banaskantha district and three weeks in Kheda. The rest of the time I stayed in Ahmedabad. The material consists of written data as well as information collected through discussions with officials in Gandhinagar and Ahmedabad. I also met officials in the towns of Palanpur, Sihori, Deesa, Radhanpur and Varahi in Banaskantha district, and in Nadiad and Balasinor in Kheda district. Villagers in 18 villages in Banaskantha and Kheda districts gave important information on the local situation.

## 6.2 THE PERIPHERY: BANASKANTHA DISTRICT

In the arid and semi-arid western periphery of Banaskantha district there was, until recently, a severe shortage of safe drinking water.

The region has scant monsoon rains, its ground water is saline, wells and ponds are brackish and most of the time they are completely dried up. It is officially classified as a 'drought-prone area'. The geo-hydrological conditions are such that sufficient water of good quality can only be obtained from distant sources such as from the Banas river. To supply the villages with sufficient safe water, distances of 100–200 km have to be bridged by pipelines. In the last 10 to 15 years the Gujarat government, initially through the Gujarat Health and Family Welfare Department, and since 1981 through the Board, has implemented several regional schemes in this area.

One of these is the Deesa scheme which supplies drinking water to 159 villages in Vav and Tharad talukas. This project is the result of the merger of eight smaller schemes. Because of an increase in salinity in the original sources, these smaller schemes were connected to new wells at the banks of Banas river, near the small town of Deesa, the headquarters of Deesa taluka where the pump station and a large overhead tank are located.

### 6.2.1 Santalpur scheme

The other large project is the recently constructed Santalpur scheme, designed to supply 72 no-source villages with water. The Board commenced its construction in 1978–79. The costs, which had amounted to Rs.75 million in 1985, were largely covered through Dutch aid.

The sources of this scheme are also located on the banks of Banas river near the small town of Sihori where boreholes, a pump station, and two large overhead tanks are situated. The distance between Sihori and the last village along the line, just on the border of the Rann of Kutch, is 100 km.

The Santalpur drinking water supply scheme is the main source of information on centralised drinking water supply systems. The organisational structure of the scheme and the Board in general, is presented in Figure 6.1.

The officials at the lowest level of the organisation are the linesmen who can best be described as watchmen. They are responsible for opening and closing the valves in the villages, maintaining the pipelines, performing small repairs, and reporting any damage and breakdown to the nearby office. They are responsible for three to four villages. Often they reside in one of these villages and receive about Rs. 350 per month as wages.

FIGURE 6.1
**Organisational Structure of Santalpur Drinking Water Supply Scheme and the Board**

Head Office Gujarat Water Supply and Sewage Board: Gandhinagar

| | | | | | |
|---|---|---|---|---|---|
| Bhavnagar Circle | Surat Circle | Rajkot Circle | Ahmedabad Circle | Palanpur Circle (Banaskantha District) Head Office: Palanpur | Baroda Circle (Kheda District) |

Radhanpur Division
(Santalpur Scheme)

Executive Engineer
Office: Radhanpur

Deputy Executive
Engineer
Office: Radhanpur

Junior Engineers Supervisors

Linesmen

To complete this organisational structure of the Santalpur scheme the following should be noted. From Sihori up to the small town of Varahi, about 60 km west of Sihori, the water flows by gravity. At Varahi the water is pumped into a water tower to supply the villages situated at a higher elevation than this small town, up to the last village along the line. Hence, at Varahi there is a pump station, where pump attendants are responsible for the electrical pump set with a diesel engine maintained as a stand-by. This station is under the authority of the Deputy Executive Engineer in Radhanpur about 17 km east of Varahi.

The project was planned and designed by the Rural Water Supply Investigation Circle of the then Gujarat State Public Health Engineering Wing. The Santalpur Water Supply scheme was approved by the state government in July 1978. It was initially planned that the 72 project villages would be connected and supplied with water within four years, in 1982. During the construction stage of the project, a Dutch expert advised on technical matters and monitored the progress of the project. The expert was an employee of a Dutch engineering consultancy firm contracted by the Department of Development Cooperation of The Netherlands.

The per capita consumption of water in the project villages has been fixed at 55 litres a day. The project has been planned to run up to 2008 and the designers used an annual growth rate of the project population of about 2 per cent per annum. Drinking water flowing from the tanks in Sihori is stored in the villages in small ground level tanks (cisterns). Through standposts, connected to the cisterns, water reaches the villagers and cattle troughs throughout the day. The main supply line is made up of asbestos (sic)—cement pipes and the branch lines to the villages are made of PVC.

The decisions on finance and design had been taken at the top administrative levels in Delhi and The Hague (The Netherlands). The planning is carried out in Gandhinagar and Mehsana, the district headquarters of Mehsana district bordering Banaskantha. The Dutch Department of Development Cooperation, through its consultants, also had a say in the planning and design of the scheme.

Officials at different administrative levels decided on the construction of the scheme. These decisions were taken at Gandhinagar, Palanpur, the district headquarters, and on the spot in Radhanpur. Components such as pipes, valves, and cast-iron joints are manufactured by private companies. The construction of the overhead tanks in Sihori, the cisterns in the villages, and the laying of pipes was done by private contractors. Tenders were invited from both companies and contractors before they were assigned the job. Transactions such as those involved in the building of large overhead tanks, and the supply of pipes and joints were decided at Gandhinagar. Smaller contracts, such as the laying of pipes and the construction of cisterns were signed at Radhanpur. The personnel of the office of the Deputy Executive Engineer in Radhanpur supervised the construction work. They also had to check the quality of the material transported by the manufacturers to the Board's depots in the project area.

The operation and maintenance of the regional supply schemes are the responsibility of the Board. In 1985 the panchayat administration had no role in these matters. In the original design of the project, the maintenance and operation were to be delegated to the Banaskantha district panchayat administration. Despite a state government resolution passed in 1973 stating that regional drinking water schemes should be maintained and repaired by the district panchayats there were no schemes where the panchayat administration was involved in Banaskantha, nor anywhere else in Gujarat. The Executive Engineer in Radhanpur and his Deputy Executive Engineers with their staff were assigned the responsibility of maintenance and operation of the scheme. The actual operation and maintenance work was carried out by pump attendants at Sihori and Varahi, supervisors and linesmen in the villages.

The Board's responsibility for construction, operation and maintenance does not mean that the local administration, namely, the village panchayats and their elected presidents (sarpanches), do not intervene in the implementation of the schemes. They were consulted, for instance, about the location of the standposts in the villages and informed the Board's personnel about the work of the contractors. Furthermore, the sarpanches tried to show and consolidate their power by interfering in the appointment of linesmen. A linesman's salary was Rs. 350 per month, a considerable amount in the area. Sarpanches pressurised the Board's officials to appoint their 'own' people. For instance, in one area the Board appointed nine linesmen who belonged to the same village and same caste as the sarpanch.

### 6.2.2 Problems in the periphery

In September 1985, the second part of the line was still under construction. Approximately 40 of the 72 villages to be supplied by the scheme had not yet been connected. In December 1983, the first part of the line between the sources at Sihori and Varahi, a distance of 60 km and connecting nearly 35 villages to the common sources, started functioning. The second part, between Varahi and Rozu (42 km), which is the last village along the line, had not yet been completed, though occasionally parts of the main line had been tested. In August 1985, water reached the small town of Santalpur, 28 km west of Varahi, for the first time during a trial run. Some experts expected that the scheme would not be completed before 1987. In any case, in 1985 there had already been a delay of three years compared with the original plans.[2]

The delay indicates how difficult it is to construct the line in an area which is the periphery of a poor district. The delay and the occasional trial runs led to the frustration of rising expectations in an area where the drinking water situation was really appalling. The villagers knew that better water would flow and in villages along the main line it did sometimes flow. One could see women waiting along the main line beyond Varahi where one day water leaked through the pipes during a trial run. Others did not wait but broke the lines, and even whole villages attacked the pipes of those villages along the main line which sporadically received water when the line was tested.

There were many leakages and pools along the first part of the line that had been put into use in 1983. These leakages and pools were due to cracks in the pipes, leaking joints and valves, and damage caused to the pipes by villagers and herdsmen wanting water for their cattle. Some villages along the first part of the line received water irregularly or not at all because the standposts were damaged or the ground level tanks in the villages remained unfinished. In other villages, untouchables complained that their women were allowed access to the common standpost only after women of other castes had finished. The waiting interfered with their household duties and other jobs.

Physical limitations, specific local conditions and incorrect policy decisions partly explain these problems. Officials in Gandhinagar and Palanpur repeatedly stressed that it is difficult to find well qualified and motivated technicians and good contractors willing to work in these remote areas. The technicians who have to live and work in the project area are 'in every respect on the wrong side', as a high official of the Board in Ahmedabad said. The area is indeed not very attractive: there are no schools, no medical facilities, frequent power failures, unsafe drinking water and social isolation. Despite these obvious difficulties, the salary of the personnel does not include a 'hardship allowance'. Only those technicians apply who have failed to find jobs in the private sector or in public bodies in urban, or less backward areas. Generally speaking, the lack of well qualified and motivated personnel in backward areas is a rule rather than an exception. Officers whose services are no longer required elsewhere are transferred to these areas. Santalpur and Radhanpur talukas are two such areas. The incompetence and unconcern of the officers are discussed in another paper (Streefkerk and Streefland [1980]).

For contractors, the area is not very attractive either. They have to work in a difficult environment and their profit margin is small. The transport, for instance, from the depots to the construction sites is not an easy task because of the condition of the secondary roads and the risk of damaging the pipes. The problem of contractors seems to be intensified by the official policy which favours local and small contractors. These contractors are given contracts to lay the lines over relatively short distances but they lack the skill and experience required to construct this type of waterworks. One of the requirements is proper levelling of the bedding of the pipes. If this is not done properly, it is difficult to fit the pipes which will eventually break. Often contractors use material of inferior quality, or leave the work without completing it because of technical or financial reasons. Poor quality of construction creates problems later on in operating and maintaining the scheme when pipes start to leak or cisterns appear to be inadequate.

Not only do the local conditions and the official policy explain the problems of construction, the intertwining of interests of high officials and private entrepreneurs in Gandhinagar and Ahmedabad is another important variable. Material of inferior quality, for example, is produced for rates lower than those which have been specified in the official contracts. The difference in price is then shared between manufacturers, sales managers, and contractors who obtained the contracts and the officials involved.

Private and public interests also intertwine at the local level, where engineers and contractors divide the spoils. One could even doubt whether the engineers are always 'in every respect on the wrong side'. The amounts involved are comparatively small, but the consequences of the public-private nexus for the proper construction and operation of the waterworks are considerable. The result is the use of inferior quality material, inexperienced contractors, faulty construction, delays, unfinished cisterns, and broken pipes.

The Board's low ranking engineers do not take decisions on tenders or contracts, but they can exert influence. For example, they are important sources of information for the contractors who compete for contracts. They know the specifications in the competing tenders. Often friends and relatives of the Board's personnel are among those who obtain contracts. In the local offices, tenders are accepted with rates that are far too low. The difference between the amount agreed upon and the official 'minimum' rates are shared between the contractors and officials. The control by the local personnel of the quality of the material supplied and the work done by the contractors also creates chances for extra earnings. Finally, the Board sometimes takes over the construction of cisterns in the project area because its personnel disapproved of the work of contractors. In this case the Board's personnel is able to make profits by using material of inferior quality and lesser quantity than officially specified. By dwelling overly on the problems of implementation I run the risk of giving a false and pessimistic picture of the situation. It should not be forgotten that the inhabitants of many of the 35 villages, connected with the scheme in 1985, had sufficient, safe drinking water almost invariably for the first time in their lives.

Specific socio-economic conditions are among the causes of the implementation problems. The question of access raised by Harijans is one of them. The other is the intertwining of public and private interests. The first can be solved rather easily by constructing more than one standpost in a village to be used by Harijans. In the original designs and reports such standposts had been provided for (Streefkerk and Streefland [1980]).

The nexus of public and private interests is a more difficult problem to cope with. It is one of the chief characteristics of the Indian socio-economic structure and is manifested in many fields of Indian society.[3] What is of interest here is whether or not centralised water supply systems suffer more from this phenomenon than decentralised provisions.

## 6.3 THE CENTRE AND ITS PERIPHERY: KHEDA DISTRICT[4]

In this section, two types of decentralised drinking water schemes will be discussed. The first type are those constructed, operated, and maintained by the village panchayats. The second consists of systems constructed by the Board and operated and maintained by the panchayat. Schemes run by the panchayat have been studied in Nadiad taluka, the inner part of Kheda district. Though the mixed type can also be found in the centre, most information has been collected in Balasinor taluka, the periphery of this rich district.

### 6.3.1 Operation and maintenance in the centre of a rich district

For several reasons one would expect that the supply of sufficient, safe water in a district like Kheda would be less problematic than in the arid and poor areas of Banaskantha's periphery. First, because Kheda district, and its inner part in particular, is anything but an arid area. Second, because of the prosperity of the area easier operation and maintenance of drinking water supply would be expected simply because there is more money available. And finally, as the provisions are in all respects far less centralised and less 'expensive', one might expect that the public–private nexus is less prominent and does not seriously hinder the proper operation and maintenance of the waterworks. Visits to villages in Nadiad taluka—three of them are situated on the periphery of Kheda's heartland and one is right in the centre—and discussions with officials of the panchayat administration and the Board in Nadiad town revealed a different picture.

The village waterworks run by the village panchayats consist of decades-old village wells with attached centrifugal pumps. Water is pumped from the wells into elevated tanks and from the tanks it is distributed through pipelines to standposts in the villages and to the houses of the well-to-do villagers. In many villages, the construction of overhead tanks and pipelines began nearly 40 years ago, 'just after independence' as explained by the elderly villagers and as indicated by the inscription on the tanks, from which it also appeared that the money had been donated by one or two wealthy villagers.

Large parts of Kheda district suffer from too much water—waterlogging—which endangers public health as it leads to mosquito-born diseases like malaria and filariasis. The incidence of these diseases has been increasing in Kheda of late, after a period of relative success in fighting both these diseases. It is difficult to determine to what extent the increase is due to the growing resistance of mosquitoes against insecticides or due to the increase in undrained water. It is almost certain that waterlogging has an important share in the increase of both diseases.

The diffusion of contaminated water in drinking water sources is another consequence of waterlogging. In several villages people complained about deterioration in the quality of water because polluted water from village ponds and other sources contaminated the village wells.

Waterlogging is a relatively recent phenomenon caused by the rapid expansion of canal irrigation and the consequent rise in the level of the water-table. It is also due to the inadequate lining of canals in an attempt to reduce the costs of canal construction. The heavy use of water for irrigation by farmers contributes to the rise of pools as well.

Another salient aspect is that there seems to be no major difference between Banaskantha and Kheda regarding water related hygiene. There are no provisions for drainage of water and standposts are in the middle of pools where cows and pigs move about and women wash utensils and clothes. These standposts provide water for the less prosperous sections of the village population. The complaints in Kheda district about the operation and maintenance of the village waterworks, however, are most striking. It appears that the supply of water is often irregular, repairs take a long time and tanks are in bad shape and often too small. These problems of operation and maintenance are closely related to the antagonism between the two major groups in the villages; the Rajputs and Patidars. Ghanshyam Shah [1975, p. 24] explains the situation as follows:

> The political clash between them is sharp and deep-rooted. It dates back to the Moghul period. The Rajput kingdoms were defeated by the Moghuls and the Patidars were appointed by the new rulers as local revenue officers and they were also invested with police authority. Later on, after the Maratha victory over the Moghuls, the Maratha conquerors continued this arrangement. Thus the Patidars accumulated power, they established supremacy over the village inhabitants. In the beginning of this century they extended their sphere of influence by being industrious and successful farmers. They acquired land from Rajputs and also colonised wastelands, the source of income of Kolis. The latter claimed Rajput or rather Kshatriya status, and looked upon Patidars as their enemies. Often Rajputs became the tenants and landless labourers of Patidars. The introduction of the tenancy acts after Independence sharpened the division between the two in the recent years. For instance, Patidar landlords evicted the tenants, largely Kolis, and turned them into hired-labourers.

At present the political–administrative structure of the villages in Kheda district and political life in the central part of Gujarat is largely determined by the Rajput–Patidar rift. The economic and political prominence of the Patidars is countered by the numerical strength of the Rajputs since the introduction of universal franchise after Independence. In the elected bodies of the village, taluka and district, and even at the state level, Rajputs have dominated. In 1985, almost two-thirds of the 96 villages in Kheda district had a Rajput village head (sarpanch), nine of the 10 talukas in Kheda district had a Rajput president, and the president of Kheda district panchayat was a Rajput too. How does the Rajput–Patidar conflict affect the operation and maintenance of the village drinking water supply facilities? The antagonism between the two often leads to mismanagement of public facilities, especially in the villages. Or stated differently, public facilities such as those for drinking water are used as an instrument in the social and political struggle between the Rajputs and Patidars—either by withholding them as the Rajputs do, or by showing that they are of little value to them, in the case of Patidars. One of the clearest examples of this phenomenon was observed in a village about 20 km from Nadiad.

In this prosperous village, with several rich Patidar families, there are about 550 Rajput and 250 Patidar houses. The remaining 250 houses belong to Harijans, tribals, Muslims, Suthars and Christians.

Drinking water is supplied from the village well. With the help of a centrifugal pump water flows into an elevated tank built nearly 30 years ago, and distributed through pipelines to six standposts and approximately 500 house connections. Five of the six standposts are located in areas where the main jatis (subcastes) reside: two are in the Rajput area, one in the Banghi falia, one in Chamar street (Banghi and Chamar are the two untouchable castes) and one in the Christian falia. The sixth standpost is in the centre of the village, near the office of the village panchayat and the talati (village accountant). The majority of the 500 house connections is in the houses of the Patidars. The Rajputs are in charge of the village with the same Rajput as sarpanch since 1974. The main problem regarding drinking water is its irregular supply. Three to four times in a year, there is no water for several weeks. This happens because the electricity bills have not been paid and the power supply has been cut off.

These problems are not characteristic of this village alone. In other villages, as well as in the offices in Nadiad town, similar complaints were heard, but as one of the informants said, 'No one seems to bother'. On the basis of the information collected in this village it is clear that neglect of, or indifference towards, responsibilities for public facilities implies more than simple disregard of obligations. The financial bankruptcy of the village and the consequent breakdown of the supply of public amenities is due to the inability of the administrators to collect water, electricity and house taxes, or because of the inefficient use of these funds. The financial mismanagement is the responsibility not only of the sarpanch, but also of the talati and other local officials. What is even more striking than this mismanagement is that, indeed, nobody seems to bother. The indifference must be explained by the enmity between the Rajputs and Patidars.

When the water supply from the village waterworks is no longer available, the villagers have to fetch water from the village well or from other sources. The Patidars, who have house connections, on the surface seem to suffer in particular. By not interfering, the Rajputs are able to show their power: 'It is their (Patidar) problem, because our women are used to carrying water', and 'why should Patidar women not carry water?'

The Patidars do not feel either deprived or thwarted. Though irregular water supply causes some inconvenience, the intermittent breakdown of the public water supply does not seem a problem to them either. If they are dependent on the supply from the well, they have servants to draw and carry the water. If they have tractors, as many do, they utilise them to carry large quantities of water from the village well. They often have wells in their fields and tractors are used to carry water from these wells to their houses. Moreover, the Patidars often have hand-pumps in their houses.

In the case of lack of street lighting, the Patidars do not really suffer. Their houses are connected with the Gujarat Electricity Board and they pay the Board directly. It is easy to light the narrow streets of the almost fortified Patidar areas in the villages by making connections with their houses.The fact that the breakdown of the facilities means a further deterioration of hygiene and a burden for the other villagers who have to walk longer distances to get water, does not seem to bother either the Rajputs or Patidars.

The material presented on the waterworks operated and maintained by the village panchayats in Nadiad taluka show that, like the centralised facilities, they too are subject to the intertwining of public and private interests. The consequences of this mix seem to be more serious for the supply of sufficient, safe drinking water, than those observed in the centralised systems. Furthermore, the nexus of private and public interests is of a different nature than that in the centralised systems. In Nadiad taluka, in panchayat operated and maintained schemes, the collusion is of a political-economic nature and its locus is the village.

### 6.3.2 Idle facilities in the centre's periphery

In Palanpur taluka, the centre of Banaskantha district, and in Balasinor taluka, the periphery of Kheda district, the geo-hydrological conditions are such that the supply of drinking water in no-source villages can be improved by providing village specific facilities. In such cases, the Board constructs provisions like wells and attached pumpsets. Once the construction has been completed, the Board leaves the operation and maintenance to the village panchayats.

In Palanpur taluka, it appeared that the panchayats are often unable to operate and maintain these facilities because of financial reasons. They have to pay pump attendants, as well as electricity bills and for repairs. In many cases the water taxes do not suffice to pay the expenses. In these villages, people either fall back on the old, unsafe and insufficient sources, or use new facilities for only one or two hours in a day.

The more detailed information collected in Balasinor taluka confirms the impressions obtained in Palanpur taluka. The office of the Deputy Engineer in charge of the construction of water-works in Balasinor taluka and the adjoining Thasara taluka is in Balasinor, headquarters of Balasinor taluka. In these two talukas, the Board financed the construction of eight schemes, of which five were not operational in 1985. The official explanation given for the panchayat's inability to operate and maintain the schemes was that the villages lacked the necessary finance to operate and maintain them. Though this explanation certainly was not without grounds, a closer look revealed another picture.

According to officials in Balasinor, the drinking water scheme financed by the Board in a village about 10 km north of Balasinor was not working in spite of the fact that the construction of the well was completed and the electromotor had been installed. After the Board officially transferred responsibility for the operation and maintenance of the scheme to the panchayat the pumpset burnt out and was not repaired because the panchayat had no funds.

Two visits to the village revealed that the situation in Nadiad can also be seen in the 'mixed' facilities in Kheda's periphery: that facilities installed by the Board but operated and maintained by the panchayat are subject to manipulation by the village elite. In this Rajput dominated village, there was a new well with a diesel pumpset attached. The construction of the well had been financed by the Board but the pumpset appeared to belong to the former sarpanch who was powerful and rich. He had been the head (pramukh) of the council which comprised four neighbouring villages. He had obtained the contract from the Board to construct the well. Water from this well was pumped by his diesel pump to irrigate his agricultural land. The people living in the area near his house and the well were provided with drinking water by the same pump using a small rubber pipe. Thus he was 'obliging' these people who were without a drinking water source of their own.

The inhabitants of this village were divided into two factions. The sarpanch of the village belonged to one of them, but was clearly no match for the former sarpanch and his followers. It is also important to note that several areas in this village had their own wells which pr vided the inhabitants of both factions with drinking water, be it less and of inferior quality.

## 6.4 CONCLUSIONS

The comparison between Banaskantha and Kheda districts was made to compare the problems of construction, operation and maintenance of, and access to drinking water facilities in the peripheral and central areas. Because the facilities in poor regions are more centralised than those in rich areas, the comparison between the periphery and the centre became one between centralised and decentralised drinking water supply systems.

Before the conclusions are presented one major methodological inconsistency must be mentioned. The centralised facilities in Banaskantha are of a very recent date, whereas those in Nadiad taluka, the centre of Kheda district are much older. I am well aware of the possibility that a future study of the piped drinking water system in Banaskantha could yield different information about the operation and maintenance of the facilities. Unfortunately, I was not able to investigate older centralised supply systems in Gujarat to verify the information collected in Banaskantha. Therefore, the conclusions I will draw have to be formulated as hypotheses.

One of the most striking results of the study is that the problems of operation and maintenance in Banaskantha district are not any more serious than those in the centre of Kheda district. The 'mixed' arrangements in the periphery of Kheda district, where the Board constructs, and the panchayats operate and maintain the village waterworks, are worse than the centralised provisions in Banaskantha's periphery. The information on the 'mixed' arrangements in Palanpur taluka, Banaskantha's centre, confirms the information collected in Kheda's periphery. In other words, centralised systems seem to be better devices for the supply of safe and sufficient drinking water than those operated and maintained by the village panchayats. Panchayats are unable to maintain and operate provisions provided by the Board either because of financial reasons, or because the functioning of the waterworks is subject to the influence of prominent villagers.

The identification of problems was one of the aims of this study. The formulation of hypotheses regarding solutions of these problems at the management level was the other objective. The intertwining of private and public interests appeared to be a main issue in both centralised and decentralised systems. However, this nexus of public and private interests is a characteristic feature of Indian society in general. So one may seriously doubt whether the problems this creates can be overcome by strategic decisions at the management level. However, the evidence presented does permit some hypotheses pertaining to management decisions. To be able to formulate them, the distinction between two types of private–public nexus must be made: administrative-economic and political-economic. The first is observed in centralised systems like the Santalpur supply system in Banaskantha. The other is manifest in the operation and maintenance of facilities undertaken by the panchayat. This distinction, and the information presented earlier, leads to the hypothesis that the economic-administrative nexus in regional schemes interferes less with the proper maintenance and operation of waterworks than does the economic-political nexus with the provision of drinking water by the panchayat.

This hypothesis can be formulated in a different manner: the implementation of public facilities, such as those for safe and sufficient drinking water, is obstructed more by the convergence of economic and political interests at the village level than by the economic-administrative nexus of more centralised provisions. If this hypothesis appears to be correct, then planners and officials in Delhi, or in Gandhinagar, should decide that the operation and maintenance of public amenities, such as the supply of drinking water, should be the responsibility of administrators less caught up in village power systems.

## NOTES

1. The concepts of centre and periphery are used here in a comparative manner. They refer to clusters of social, economic, demographic, infrastructural and physical characteristics. Henceforth, synonyms like rich and poor regions will also be used.
2. The scheme was indeed completed in 1987. It became the 'life-line' of the villagers during the severe droughts in that year.
3. Since Gunnar Myrdal introduced the concept 'soft state' in 1968 this phenomenon has been acknowledged as a feature of South Asian societies. Robert Wade [1985] writes about it in an illuminating manner for fields such as irrigation.
4. The panchayat waterworks in Nadiad taluka generally are three to four decades old, so they are not new technologies in the strict sense. They are included in this study because they are village specific facilities and provide information on the problems of operation and maintenance of such systems run by the village panchayat. But more important is that they constitute the drinking water supply in a central area, so they have to be included for methodological reasons. Information on the waterworks in Palanpur taluka, the centre of the periphery, is presented in the part dealing with Balasinor. This is done as there are similarities between the problems faced by the villages in Palanpur and Balasinor talukas.

# 7

# A CONSTANT FLUX OF PERSONNEL: SOME IDEAS ON PERSONNEL TRANSFERS IN BUREAUCRACY

**Frank de Zwart**

> **Transfer from one place to another is that part of his life that a civil servant dislikes most and getting a government servant transferred to either secure his gratefulness or to penalise him is the most popular occupation of politicians (Mohan Mukerji [1982]).**

## 7.1 INTRODUCTION

This chapter examines how nurses, clerks, chemists and doctors, working in the Gujarat Public Health Department, cope with the system of frequent personnel transfers to which they are subject. I argue that this system forces them to adopt strategies to cope with transfers, that end up defeating the system's official aims.[1]

For government, transfers or the threat of transfers have several advantages: the possibility of an unwanted transfer may keep civil servants from violating bureaucratic norms; frequent transfers prevent civil servants from becoming entangled in local affairs, since such entanglement could endanger their supposed impartiality; and, finally, transfers enable government to use expertise economically as personnel with special skills can be employed wherever they are most needed.[2]

For civil servants, however, transfers have another meaning. First of all, transfers have a profound, and possibly disastrous, impact on their private lives. Second, civil servants do not believe that organisational advantages are the reason for the transfer, but suspect corrupt[3] transactions behind it. In order to influence decisions on their transfers, civil servants seek access to the networks in which they think these transactions are made. To gain such access, however, requires time and money, and this is inevitably at the expense of their duties.

I will not deal with the official reasons for transfers and their possible advantages for government. Instead I will try to describe the transfer system as the civil servants see it. Thus following Herbert Simon's advice that

> (i)n the study of organisation. . . the operative employee must be at the focus of attention, for the success of the structure will be judged by his performance within it. Insight into the structure and function of an organization can but be gained by analyzing the manner in which the decisions and behaviour of such employees are influenced within and by the organization (Simon [1976], p. 3).

## 7.2 MOTIVES FOR ACTION

Transfers, and all that has to do with them, are a common topic of conversation among civil servants. Colleagues discuss each others' transfer histories and feelings for and against various locations. They also like to exchange notes on success and failure in getting the transfers of their choice. From such observations it becomes clear that civil servants do not consider themselves at the mercy of impersonal bureaucratic decisions taken at the top. They believe that with some effort and the right strategy one can do a lot to prevent disaster: a transfer to a place one really dislikes. This is a matter of great concern to civil servants. They can all be transferred at any moment[4] and have very strong feelings for and against certain places. Clerks, nurses and compounders, class III personnel in civil service jargon, consider themselves successful if they have a posting in or near their home town. They prefer that to any other since they like to work in a familiar environment, within close reach of friends and relatives. Moreover, a home posting means that they do not have to spend money travelling to and from their work place during weekends and holidays.[5]

Doctors are often prepared to go further afield. Unlike class III servants, who prefer a particular place, doctors prefer a certain kind of place: where they can have a profitable private practice, or send their children to a good school, or where they can just lead a comfortable life. (It will be seen that, to a certain extent, these preferences are mutually exclusive.)

It is illegal for government doctors to have a private practice, i.e., to charge their patients. Many doctors, however, ignore this rule, especially if they are posted away from their home town. In their home town they would not charge their patients since, as one doctor explained, 'you cannot exploit your own people'. Moreover, a private practice would not be very profitable. A doctor who works in his home town or village has to oblige even more people than a class III public servant since, by the nature of his trade, he has more to offer. On the other hand, in a place where he has no relatives and only a few friends, there would be few people to oblige, but there would be many trying to oblige him with presents or cash.

There is a general notion among people that a government doctor with whom they have no personal ties, will not treat them well. Some doctors complain about having to refuse people who insist on paying them. Others maintain that they have stopped refusing and now accept token money in order to be trusted as capable physicians. Obviously, if a doctor wants to make money, he is better off in a place where he has few personal ties with the people. For this reason many doctors are not bothered too much by frequent transfers.

Another reason for this attitude is that for most doctors a government job is not an ultimate goal. Usually they resign after a few years and start their own practice. The best time to do this is after having saved enough money to invest in a private practice. For some doctors an unfavourable transfer is reason enough to resign no matter how much they have saved, but many are really determined to make a lot of money during their tenure of government service. They try to be posted to a good 'earning centre', and are willing to pay for that.

A good earning centre has a number of qualities. There should not be too many relatives. As explained earlier, they consume a doctor's time and cannot be charged for it. There should also not be many private practitioners. Since people who can afford it would consult them as, on the whole, government doctors have a bad reputation (Van der Veen [1981]). Moreover, private practitioners often complain to the authorities if they feel that a government doctor undercuts their business. Therefore, cities, with their numerous legal private practitioners, do not rate high as earning centres.

Another feature of a good earning centre is its location in the network of roads. Health centres near main roads are frequented by controlling officers on tour, who usually stay close to the main roads because they do not like to travel through jungle areas. For a government doctor who charges his patients, inspection can mean several things: if a strict inspector finds out, he will punish the doctor, most likely with an unwanted transfer. But if the inspector is not too strict, he may demand part of the profit in exchange for his permissiveness. In any case doctors prefer not to have inspectors around, and good earning centres are thus located away from the main road.

In order to continue their practice undisturbed, doctors should be on good terms with the local leaders. It does not, of course, go unnoticed if a government doctor practices in a village. Good relations with the local leaders may help avert complaints and unwanted transfers. A doctor who wants to continue in a posting needs political protection. But, at the same time, he is well advised to try not to become too closely involved in local politics. Otherwise, he risks being transferred as soon as the political climate changes. One informant was transferred after his political patron had been thrown out of power during an election. The new ruling faction preferred a doctor who could be counted on to give them his support. Therefore, the doctor was transferred at their behest, to what he calls 'a very bad place'. His successor, he complained, could simply take over his practice, whereas he had to start all over again as the new place had no 'tradition'. This term is often used to indicate the habits of a predecessor.

Maximising their income is, of course, not the only aim doctors have. Really good earning centres, where a doctor can make about Rs. 5,000 per month in addition to his salary, have to be bought. It would cost an initial sum of Rs. 10,000 to Rs. 20,000 to be posted there in the first place, and often a part of the profit, in the form of an annual percentage, to keep the post. Some doctors, even though they may plan to have an illegal private practice, do not want to get involved in this arrangement. They settle for a less remunerative place; one they would not have to buy. Moreover, good earning centres may be undesirable places to live in. Interior areas, for example, have the required characteristics of a good earning centre, but are also disliked because they lack facilities like good housing, English language schools, clean food and water, and, in general, cultural life[6] (see also Gupta [1981]).

If a doctor has children this becomes a real problem. He may then settle for les money in order to have a better quality of life. Discussing this topic, one doctor remarked: 'What good am I as a father if I earn in a place where I cannot even put my children in a proper school?' His colleague commented: 'It is no use earning in a place where you cannot even spend the money'.

Thus, for most doctors, a post in the interior is only attractive for some time, usually before the children reach school-going age, or after they have left school. Contrary to class III servants, who almost without exception, want to work in their home towns, doctors preferences often change during the course of their career.

## 7.3 THE PUBLIC INTEREST AND OTHER REASONS FOR TRANSFERS ACCORDING TO CIVIL SERVANTS

Most civil servants do not discuss their feelings for and against places openly. They feel manipulated by the transfer system, and fear that revealing their personal preferences makes them an easy target for the authorities. For the same reason, they also hesitate to protest against unwanted transfers, however unreasonable they may think the transfers are. A doctor who had been working in the same place for the last seven years explained that even though he was proud of the good work he had done in this area, and he and his family were quite happy there, he would not protest if he were transferred. If he protested, he said, his superiors would think: 'Why is this man so eager to remain in this place; he must have a profitable business there'. If he protested he could be suspected of corruption, and lead the authorities to demand a share in the profit they presume he makes. In fact, he said, this racket could be the very reason for their trying to transfer him in the first place.

This doctor's ideas are not exceptional. The concept of the 'public interest' as a motive for transfers is commonly distrusted. Distrust that is constantly reinforced by the unclear and unexplained procedures which the government adopts to transfer people. Individual transfers are not explained. Civil servants are simply ordered to proceed to another location 'in the public interest'. Moreover, the frequency of transfer varies greatly from person to person. Some manage to remain in one place for over 15 years; others are transferred 10 times during that period. These differences also remain unexplained, which does not help convince civil servants of a public interest behind transfers. To them, transfers need further explanation, and they are never short of their own theories about them.

## 7.4 HOW TO OBTAIN OR EVADE TRANSFERS: THE PRESUMED NEED FOR MONEY OR INFLUENCE

In the theories of civil servants, the public interest is not taken very seriously. Transfers, according to them, are a result of corrupt financial transactions, or simply of favouritism.

Those who have no money, but desire a favourable transfer, need influence. Influence can be any kind of relationship with a powerful person. It need not necessarily be a close one; to know a friend of the brother of the personal assistant of a deputy director, for instance, would certainly be considered influence.

Influence (*laghvagh* in Gujarati) is a very wide concept. Let us consider an example. The nephew of a Member of Legislative Assembly (MLA) had always worked where he liked. His uncle helped him, so he was not worried about unwanted transfers. He himself occasionally helped a friend, but not too often, as he said, because that would upset his uncle. But for his friends, just to know him means to have influence.

Another informant managed to be posted to his home village simply by taking the trouble to visit the Directorate Office in Ahmedabad three times. When he had finally succeeded in convincing the right clerk of his case, he was transferred within no time. He called this influence too.

A health visitor, who was very eager to be transferred back to his home town, waited for his chance to meet the Deputy Director of his sub-division who would visit the centre where he was posted on one of her regular tours, and when she did he tried to ingratiate himself. He explained to his colleagues that she sometimes exchanged favourable transfers for attentive behaviour, and this information he had obtained from colleagues in a former posting. But his attempt was not successful. She did not appreciate his approach, and he did not get his transfer. His colleagues were convinced that he had ruined his influence, and with it his chances of being transferred in the near future.

Obviously, since it is so wide a concept, a great many people claim to have influence. Many more, in fact, than the number of people who actually work in the place of their choice, which suggests that influence is not as effective in getting transfers as people believe it to be. Nevertheless, few people deny that influence works. One reason may be that to have or to use influence gives status. To deny having or using it is to deny oneself status. One informant had, with the help of his friends, approached the MLA of his district, who had promised to help him with his transfer. He also sent a formal request for a transfer to the Directorate of Health Services in Ahmedabad. After two months he got his transfer, and he boasted about his influence and praised the MLA who had helped him. The MLA, however, had contacted the district officer, the informant's immediate superior, and told him not to worry about losing his staff this way. He had not done anything in this case, but the officer should understand that as an MLA he could not say that to his voters. He would always promise to help them and claim credit for success later as that would enhance his image as a capable politician. The officer was asked not to talk about this to his colleagues, which would be unlikely as it is not wise to offend an MLA.

Another reason why the effectiveness of influence cannot be easily denied is that, like magic, influence can explain both success and failure. Someone who wants to be transferred usually begins by sending an official request, and then proceeds to back this up by exerting influence as a request is considered useless without it. If he gets the requested transfer, he will always claim that his influence worked; but if he does not get the transfer he will blame it on the better influence of another applicant for the same post. Claiming influence is, of course, easier for people who were never transferred away from the place of their choice. They can simply say that they are protected by a powerful patron.

## 7.5 HOW STRATEGIES FOR OBTAINING AND EVADING TRANSFERS AFFECT PUBLIC HEALTH CARE

Popular ideas about the transfer system fit Myrdal's [1970, pp. 232–33] concept of 'a folklore of corruption'. The ideas are perhaps not always accurate, but they are very real in their consequences since people act upon them. The strategies that class III servants adopt in order to get transfers are based on the assumption or hope that influence works. Whether or not this assumption is correct, their strategies have profound and overall negative impact on the performance of the organisation as a whole.

It is not always easy to make influence work. Politicians seem to spend a lot of time supervising transfers; and civil servants certainly spend a lot of time trying to get them. Influence entails time consuming social visits, attempts to be introduced to important people, waiting for appointments with busy officials, etc. It thus requires days off from work. In most health centres I found staff members on leave for a few days, sometimes even weeks, arranging their transfers in the capital city. The same has been observed in other areas (Justice [1983]; Wade [1982]), and in Gujarat it is a common complaint of officers, since work in their centres is often hampered because staff members are absent.

Another strategy that civil servants adopt has the same effect. If someone is transferred to a place he dislikes, one possible way out is not to commence duty in this new post, but proceed on leave without pay. The advantage here is that this gives plenty of time to lobby for a better transfer, and after some time the officer in charge will seek a replacement in order to fill the vacancy, which, if he succeeds, improves the chances of the first man getting another transfer. This strategy has the obvious disadvantage that one cannot be sure that the new transfer will be more desirable. Nevertheless, there were vacancies caused by this strategy in one-third of the health centres I visited. For public health care this is quite disturbing. Especially as there is a high degree of specialisation among health personnel, so that an absent staff member may render a hospital without an X-ray facility for some time, or a district without health education.

One of the reasons government resorts to transfer is to penalise poor work, but I doubt whether this is very useful.[7] Civil servants who are transferred against their will may also refuse to work, even if they do not proceed on leave without pay. Most civil servants see no clear relation between the way they perform their job and their chances for obtaining desired transfers. Hard work is not, in their experience, the decisive factor in getting the transfer of their choice. Desired transfers are the result of pulling strings and not hard work.

The problem is most serious in backward, tribal areas that are used for punishment. Compared with other areas, punishment areas have more people who are extremely dissatisfied. Thus, transfers have a greater impact on local bureaucratic performance. Dharampur is an example of a notorious punishment area. Gupta [1981], evaluating a development project in this area, writes that implementation was seriously hampered by 'poor morale and high turnover' of civil servants. The efficacy of the coordinating committee, of which Gupta was a member, 'was seriously limited by the high rate at which its members were transferred. Within two years . . . it saw six District Collectors, five District Development Officers, and three Project-cum-Tribal Development Officers. Each change required that the new officer be briefed on project purposes and findings, and his collaboration solicited' (Gupta [1981], p. 116). Obviously, people are eager to move out of Dharampur and they often succeed. A side-effect of high staff turnover is that it demands excessive filing. The elaborate bureaucratic filing culture, 'the bureaucratic machinery . . . pushing files more than pushing the frontiers of development', as Gupta [1981, p. 121] puts it, was introduced by the British partly as a result of their transfer policy. Rapid turnover demanded that new officers be informed about the area and cases they had to deal with, and the elaborate filing and noting system, a well-known source of delay, developed from it (cf. Heginbotham [1975]).

High turnover of government doctors leads to loss of patients. Patients prefer a doctor they know, so that the clientele of a doctor grows if he remains in one place (Djurfeldt and Lindberg [1975]). Hence, if a doctor is transferred, his successor starts with 20 to 40 per cent fewer patients. A loss that is only gradually recouped if and as the new doctor becomes known in the area. But that is the dilemma as frequent transfers are intended to prevent just that. A doctor who becomes known in an area develops friendships and enmities which may interfere with his practice according to universally accepted standards. Therefore, bureaucrats and politicians often claim that frequent transfers prevent corruption, which is probably true for what James Scott calls parochial corruption, 'where ties of kinship, affection, caste, and so forth determine access to the favours of power holders' [1972, p. 88]. It is not true for what Scott calls market corruption, 'which provides access to those who can pay for it' (*ibid.*). On the contrary, in the Public Health Department, transfers promote market corruption. As has been noted, patients feel a strong need to oblige a doctor. They feel that a doctor who is in one way or another obliged to them will treat them better. If they cannot oblige him by stressing a personal relationship, they will oblige him with money.

Frequently transferred doctors have few personal and affective ties with their patients, while patients without ties to their doctor feel the need to pay him. This may be beneficial for the doctors' private practice, but not for public health care which is supposed to be free of charge.

## 7.6 CONCLUSION

If, as government claims, fighting corruption and promoting efficiency are the reasons for transfers, it seems that the system in its current form defeats these very objectives. Transfers, although in theory perhaps a good management tool, are counter-productive. Civil servants feel manipulated and discriminated against when transferred without a proper explanation. They perceive a transfer as a threat and try to insure themselves against it by bribes and other illegal practices. After all, according to their perception theory, that is how the system works.

It would be better for civil servants themselves and for public health care to make transfers more predictable. It could, for instance, be laid down in advance, when and to where people shall be transferred. If civil servants were a bit more certain about their prospects, they would not have to spend so much time trying to insure themselves against possible disaster.

Moreover, if only those transfers are required which can be explained in terms of concrete administrative needs, instead of general statements like 'in the public interest', the number of transfers would probably decrease, and so would the number of cases for which people see but one valid explanation: corruption. A way out of self-fulfilling prophecies is needed.

In India, with its vast differences between regions, some transfers are probably unavoidable; if only to fill posts in unpopular areas. However, the situation now exists where one civil servant does all he can to get a posting in his native area, where all posts appear to be taken by other civil servants who, in their turn, also try to be transferred to their native area. They spend a lot of time and energy at the expense of their work.

## NOTES

1. The distinction between official and unofficial aims is important for explaining the persistence of the transfer system. Elsewhere [1985] I have distinguished manifest and latent functions of transfers. Manifest functions being beneficial to the system, the bureaucracy, as a whole; latent functions benefit individuals. A detailed case study by Robert Wade [1982, 1985], dealing with an irrigation department in South India, shows that the transfer system can be an effective instrument for politicians and top officials who aim to build up political power or to become rich. In fact, Wade suggests that these unofficial aims, latent functions, explain why the transfer system persists even though it fails to fulfil its manifest function: to combat corruption and promote efficiency.
2. These advantages probably led Pai Panandiker, a prominent scholar on development administration, to note that: 'In the Government of India the policies on transfers are certainly one of the better features of its personnel administration' [1966]. This observation does not, however, refer to the real world. It deals with an organisational form and its eventual possibilities. Analyses at the level that Herbert Simon proposes leads to a different view on transfer policies. Pai Panandiker is obviously aware of that since he adds to the above notion that, '. . . in operational terms the practices had somewhat negative overtones' [*ibid.*]. He does not, however, elaborate this point.
3. Wertheim [1963, p. 131] defines corruption as '. . . the conduct of officials who infringe the principle of keeping their public and private concerns strictly separate'. Wertheim argues that in the recently independent states in Asia, this principle was not generally accepted by society as a criterion for proper conduct on the part of civil servants [*ibid.*]. But today, 25 years later, Wertheim's definition is quite generally accepted and is frequently used in India. Indians now consider corruption as reprehensible as Europeans do.
4. Transfer histories of 60 informants show that, on the average, doctors as well as class III personnel are transferred once in every 5.5 years. Note, however, that an important characteristic of the transfer system is lost in this figure: transfer frequencies differ strongly between individuals.
5. A home posting can also be disadvantageous: being near friends and relatives means having to 'oblige' them. A relative with a government job is seen as a valuable resource to be exploited, so that a civil servant who works in his home town gets many visitors who ask for special favours like free medicine, quick treatment, and an introduction to the doctor. This can be a nuisance, especially if

superiors complain about it while, at the same time, it is of course hard to refuse relatives and friends. For this reason some civil servants seek a place nearby, but not in their home town. Here they can both do their job and be useful to their friends.

6. On the whole, doctors are more metropolitan oriented than class III servants. Not that they only wish to work in cities, as the stereotype goes, but they do become more easily dissatisfied and bored with village life.
7. In tact transfer is the most workable penalty government has. Indian civil servants are very well protected against other possible sanctions like dismissal, reduction in rank, or delay of promotion. According to Pai Panandiker, this 'anxiety to protect the civil servant . . . indicate(s) a general distrust about the efficacy of a single hierarchy . . . . In India fear of linguistic, religious and caste biases even of higher officials as well as fears of political implications of any significant staff retrenchment have brought . . . many safeguards to aid civil servants . . . and lack of an effective procedure for prompt disciplinary actions' [1982, pp. 196, 197, 198]. Transfers are an exception to this.

# II. ENERGY PROGRAMMES

# 8

# ENERGY PROBLEMS: AN INTRODUCTION

The first Indian coal mine was opened in Bengal in 1820. After 30 years, in 1854, three other mines were started. 'In that year the commencement of the East Indian Railway line, which was laid to run through the coal-bearing regions in the Damuda basin, gave an impetus to the mining industry and new pits were opened in large numbers' (Gadgil [1972], p. 58). Initially, the expansion of the coal mining industry was related to the development of Indian railways. After 1900, modest industrial development also contributed to the development of the mining industry. On the eve of Independence it was an important industry and coal was the main commercial source of energy. However, despite the availability of coal in South Asia and the development of the mining industry, India had to import coal. This was because of the unfavourable location of the coal fields in the eastern part of the country—in Bengal, Bihar and Andhra Pradesh. The majority of coal imports went to the Bombay Presidency (Gadgil [1972], p. 59).

For the same reason the consumption of oil increased rapidly after 1918. It was used as a substitute for coal on the Western coast and was imported from Persia because of its proximity to Bombay. The cotton textile industry was one of its most important consumers (Dasgupta [1971], p. 30).

Kerosene was the main oil product consumed in India chiefly for illumination purposes. It came from Burma, where oil had been discovered around 1890. Burma remained India's most important source of oil between 1905 and 1942 (Dasgupta [1971], p. 17). In the twenties, the share of kerosene in the total consumption of oil fell from 60 per cent to 56 per cent. This decline was due to the growing consumption of motor spirits as well as industrial expansion. Burma became the almost exclusive supplier of motor spirits, while Persia remained the main supplier of fuel oil until 1951 (Dasgupta [1971], p. 25).

During the first 15 years after independence, the level of consumption of commercial sources of energy was still comparatively low. Dasgupta [1971, p. 91] writes that, according to estimates for the sixties, non-commercial fuels, such as wood, cow dung and leaves, accounted for 68 per cent to 78 per cent of the total Indian energy consumption. These percentages, however, do not indicate drastic changes in the pattern of energy consumption since the early fifties, and particularly since the sixties. Between 1953 and 1979, the consumption of commercial sources of energy rose by 390 per cent. The consumption more than doubled between 1953 and 1966, and again doubled between 1966 and 1979 (United Nations Conference [1981], pp. 5, 8). Even more important are the changing ratios of the consumption of commercial and non-commercial fuels since 1965—commercial sources of energy began to dominate the total consumption of energy. From 1965 onwards the proportion of non-commercial fuels in the total energy consumption decreased from 55 per cent[1] in 1965 to 44 per cent in 1973 (Bapna [1985], p. 21).

The altered pattern of energy consumption reveals important changes in the Indian socio-economic structure, such as industrial modernisation and growth; the electrification of villages; the 'Green Revolution' and the fuel required to run tractors and pumps; the increased use of private cars and trucks; and the emergence of an 'energy culture' among the urban and rural elites. This stimulates the demand for cooking gas, electricity for illumination and maintenance in houses (Bapna [1985], p. 20).

At present India faces two serious energy problems. The first problem is caused by the constraints on commercial sources of energy and the supply of oil in particular. The second problem is the scarcity of non-commercial sources of energy in rural areas.

In the seventies and eighties, coal, electricity and oil were the main sources of commercial energy. Coal is still the most abundant resource available in India. In the eighties it contributed almost 38 per cent of the total commercial energy supply (United Nations Conference [1981], p. 9). It is a main source of energy in all sectors of the economy, apart from agriculture. However, the problem of coal remains one of transportation because of the distant location of the coal fields. Furthermore, it will be difficult to increase the production of coal to levels required by future demand (United Nations Conference [1981], pp. 7, 9).

Oil accounted for 10 per cent and electricity for more than 50 per cent of the total energy consumption. These statistics[2] must be treated carefully because the figures on coal and oil refer only to quantities directly consumed for energy uses. They do not include coal and oil used to generate electricity.

The share of oil is the smallest of the three commercial fuels, but it is the most expensive source. Oil imports are an enormous burden on the Indian balance of payments. Oil is not only a main source of energy but an important raw material in the production of petro-chemicals and fertilisers. In the late seventies, about 40 per cent of the crude oil requirements were met through domestic production. The remaining 60 per cent had to be imported, absorbing about two-thirds of India's foreign exchange (United Nations Conference [1981], p. 7)

According to Moulik [1982b], the distribution of the consumption of the three sources of energy in rural areas in the seventies was: oil 50 per cent, coal 25 per cent and electricity 25 per cent. During the same period, transport accounted for 50 per cent of the oil consumption, domestic use (kerosene) 30 per cent, industry 10 per cent, and the share of agriculture in the consumption of oil was 5 per cent. Only 12 per cent of the total production of electricity was consumed in rural areas (Moulik [1982b]).

The use of modern sources of energy is, like modern health care, unequally distributed between urban and rural areas and between rich and poor. The urban well-to-do and the village elites are the main consumers of commercial sources of energy.

The energy needs of India's rural population constitute the second energy problem. The crisis of the commercial sources of energy hardly touches the majority of the villagers, except insofar as it leads to higher prices of a few mass consumption items. More than 85 per cent of the energy required in the villages is for cooking (Moulik, Shukla and Sinha [1985], p. 4). Most of the villagers cannot afford to use commercial fuels. They survive on sources such as firewood, dung-cakes and agricultural wastes. Firewood is the most important source. It is estimated that out of these three sources wood accounts for the largest share with almost 65 per cent. These sources are supplemented marginally by commercial fuels like kerosene. Other commercial sources are almost exclusively used by a small affluent section of the population (Moulik [1982b]).

The doubling of the Indian population during the last 30 years has led to a tremendous rise in the demand for kitchen fuel. At the same time, the sources of firewood have been decimated in many parts of the country (Hooja [1985], p. 38). The growing commercial demand for timber has led to deforestation. Elsewhere, common lands and wastelands have been converted into agricultural land, or its vegetative matter has been monopolised by the panchayats for commercial purposes. Scarcity of firewood and its commercialisation have forced children and women to cover larger distances in search of wood fuel. They have to walk for miles to find bushes, branches or twigs which they then carry home. Last, but not least, the scarcity of fuel adds to the problems of proper nourishment. Often the little amounts of food available cannot be prepared because there is no kitchen fire.

The squeeze on the use of conventional fuels and the growing shortage of non-commercial sources of energy has forced the government to increase the utilisation of nuclear energy sources and hydro power.

Hydro power is a major source of renewable energy. Wind, solar energy and biogas are other types of renewable energy which the government is trying to develop. These sources are meant to meet the requirements for lighting, heating, pumping and low grade heat in the countryside.

Interest in renewable energy dates back to the start of the Gandhian movement when it was included among the ideals of independence, village industries and village autonomy. Today the development of renewable sources of energy is an integral part of official plans. In this respect India is exceptional compared with other Third World countries. The promotion of alternative sources of energy has strong political support in Delhi and is well established in the national government administration. In 1982, the Department of Non-Conventional Energy Sources (DNES) was set up in Delhi with the intensification of research and development activities as one of its objectives. The Central Commission for Additional Sources of Energy (CASE) is another important national government body.

At the state level the propagation of the use of non-conventional energy also enjoys considerable political and administrative support. In Gujarat, for instance, the Gujarat Energy Development Agency (GEDA) plays a prominent role. Semi-autonomous institutions like the Gujarat Agro Industries Corporation and the Khadi and Village Industries Board also contribute to the promotion and implementation of alternative sources of energy by providing prospective users with financial support. Apart from government institutions, many private industrial firms and several NGOs are involved in the production and implementation of biogas plants, improved stoves, windmills and solar energy systems. The results of public and private endeavours to spread the use of non-conventional sources did not remain restricted to a few isolated R & D projects or demonstration centres. For instance, in 1985, there were 78,512 biogas plants in India, of which 19,753 (25 per cent) were located in Maharashtra and 4,468 (almost 6 per cent) in Gujarat (Annual Report [1985], p. 9).

Biogas plants were introduced as decentralised and self-reliant energy systems in the thirties. They were also intended to make villagers more aware and to stimulate their political autonomy.

The improvement of rural living standards remains a goal of renewable energy programmes. However, the family size biogas plants are not meant for the majority of the rural population as they require cowdung from at least three cows and several thousands of rupees to finance the plant. At best, they are only suitable for the cattle owning upper income stratum of the rural population which accounts for no more than 8 to 10 per cent of the rural population. Research conducted in Gujarat revealed that, in the early seventies, the average landholding size of the biogas plant owners was as high as 26 acres (Moulik [1982b]).

The diminishing accessibility of poor people to cowdung for use as fuel in villages with biogas plants is another problem.

Despite its elite bias, biogas can contribute to the solution of rural energy problems on a larger scale, provided that community size biogas plants are set up. Biogas is a multi-output technology producing energy, manure and carbon dioxide. These products can be used for cooking and to provide power for village street lighting while the projects also generate income (Moulik [1982b]).

Blankenberg provides a detailed analysis of the national biogas development project in Gujarat and Tamil Nadu from the perspective of strategic management. One of his main conclusions is that there is a lack of management tools, for example, the analysis of needs or monitoring during the process of strategy development.

Ahuja *et al.* also studied the implementation of biogas programmes in Rajasthan. They clearly highlight the differences between general macro-assessments and specific local circumstances and needs. They advocate active participation by the beneficiaries in the programmes in order to bridge this gap. They also note that farmers make realistic comparisons between the costs and benefits of traditional sources of fuel and those of the new biogas technology. This may facilitate the farmers' future active participation in energy programmes.

The installation of improved stoves in the villages is another means of coping with the rural energy crisis. If well constructed and maintained, their energy efficiency is considerably higher than that of traditional stoves. They also contribute to improving the health of village women as they produce less smoke and may therefore reduce the incidence of respiratory disorders. The chances of successfully implementing these programmes decrease when only men are responsible for them. In India, men and women traditionally lead largely separate lives in their daily activities; the relations between them are hierarchical with the domestic sphere being the female domain. Hierarchy is even more significant when poor women are the main beneficiaries of a project as they lack the social status that might counteract the status of men.

Schenk-Sandbergen shows how important it is for women to participate if the stove programmes are to be successful. Women have much easier access to the female beneficiaries in the villages who must be motivated to use the new technology. But Schenk-Sandbergen's data indicate that highly motivated men can also inspire success.

Skutsch, the last author in this section, studied the introduction of improved stoves and the related social forestry programmes from another perspective. Her focus is on the management practices of the NGOs involved and on the criteria for success or failure of projects. She argues that NGOs are interest groups whose criteria for success are such that programmes can be considered successful at different levels for different reasons.

## NOTES

1. This percentage is indeed lower than the figures presented by Dasgupta for the same period.
2. The figures are based on estimates. They represent a situation where no deliberate measures were introduced to manage the energy supply and demand.

# 9

# STRATEGIC MANAGEMENT OF THE NATIONAL BIOGAS DEVELOPMENT PROJECT IN GUJARAT AND TAMIL NADU

**Floris Blankenberg**

## 9.1 INTRODUCTION

In 1981, a large-scale National Project on Biogas Development was launched by the Government of India, with the aim of installing 400,000 family size biogas plants before the end of the Sixth Five-Year Plan period. According to this Plan, 35,000 biogas plants were to be distributed in both Gujarat and Tamil Nadu. The most recent estimate of the total number of family size biogas plants installed in India, including approximately 100,000 plants that were introduced before the National Project was launched, is one million.[1]

In this chapter, factors which play an important role in the strategic management of this project are identified and described. The reason for selecting the National Project on Biogas Development as a subject of study is that among various programmes of renewable sources of energy undertaken by the Indian government, the biogas project is an important and promising one. Much can be learnt from the experiences gained in this project.

This contribution has no pretension of being very analytical. It should be considered as an illustration of some management problems that occur in the process of implementation of rural technology projects in India. An analysis of the National Biogas Development project was placed in a much broader theoretical context in a previous publication (Blankenberg [1985]).

Two states of India, Gujarat and Tamil Nadu, are included in the review. There are some remarkable differences in the management structures between these states which offer a good opportunity to compare them and draw conclusions about the effectiveness of the different management approaches.

Some of the concepts that were introduced by Chaudhury (1985) will be used as tools to guide the analysis. The authors distinguished certain variables and detailed them as follows:

1. The strategy underlying the project, referring to the long-term choices concerning a programme such as the ideology, goals, the design of services, action plans, target groups, and organisational resources and capabilities.
2. The process of strategy development, referring to the role of influential persons and important organisations involved in a programme and the use of management tools such as needs analysis, generation of available alternatives and collection of information required for decision-making.
3. The process of strategy implementation, referring to the nature and role of implementing organisations, to intra- and inter-organisational coordination, to the staffing position of implementing organisations, and to monitoring and evaluation.

After a short review of the project, the above concepts will be applied to the situation in the two biogas projects. Each of the variables will be briefly discussed to give an idea of the nature of strategic management. Summarising remarks are added to each section and in the last section some cautious conclusions about the management factors that determine the success of biogas plants are formulated.

## 9.2 HISTORY OF THE BIOGAS PROJECT

The National Project on Biogas Development could be considered as a follow-up programme as it amalgamated a number of smaller experimental projects which existed at the local and regional levels. Before 1981, small non-governmental institutions distributed biogas plants in some states. In 1981, the Government of India became deeply involved in the scheme, distributing biogas plants in all states on a large scale. Research and development activities in biogas technology had been initiated in the thirties by various institutions, especially the Khadi and Village Industries Commission (KVIC). Small institutions at the local level also experimented with biogas technology.

From 1960 onwards, the KVIC became involved in the provision of subsidies, loans and technical guidance, training of masons and extension work and in the development of improved designs for gas burners, lamps, and other appliances. The Government of India participated in the KVIC programme from 1974 onwards by providing subsidies. Prior to 1974, 7,000 biogas plants had been installed in India but the number increased to almost 100,000 by 1981. The role of the KVIC and a number of local institutions has been of crucial importance for the development and initial dissemination of biogas plants.

The project received a major impetus in 1981 when the national government decided to extend its full support. The measures taken under the National Project resulted in a sudden increase in the number of biogas plants installed.

## 9.3 STRATEGIES UNDERLYING THE PROJECT

In this section the long-term choices of the national government concerning the biogas project are reviewed.

### 9.3.1 Ideologies

The Indian government became deeply involved in the project in order to promote the following objectives: (*a*) to reduce increasing pressure on the demand for firewood and commercial fuels, (*b*) to conserve and utilise organic materials as high quality fertilisers in order to supplement and optimise the use of chemical fertilisers, and (*c*) to utilise and extend experiences of institutions, often based on Gandhian principles, in the introduction of decentralised and self-reliant energy systems at the village level.

### 9.3.2 Goals

The formulation of official goals was in line with the ideologies mentioned earlier: scarce sources of energy had to be saved and organic materials had to be utilised as fertilisers with the help of decentralised technology inputs. The number of biogas plants to be installed under the project was fixed at 400,000. It can be added that bureaucratic institutions within the government administration were eager to participate in this rapidly expanding project.

### 9.3.3 Design of services

The strategy concerning the design of services to be delivered was characterised by detailed decision-making at the national level. Aspects that were to be included in the service delivery system were direct financial support for beneficiaries and implementing agencies; the involvement of nationalised banks in credit schemes; financial support for the organisational arrangements of the various state governments; facilities for the set-up of monitoring and evaluation cells at the state level; financial support for research and development activities; and facilities to make key construction materials available.

### 9.3.4 Action plans

The translation of the design of a service delivery scheme into an action plan was to be the responsibility of the state governments. Yet, as stated earlier, the directives issued at the national level were very detailed. Action plans were, therefore, not much more than elaborations of details; there were only a few opportunities to introduce variations on the basis of state or region specific circumstances or needs.

### 9.3.5 Target group approach

First, a number of biogas plants for each state was fixed. Second, it was decided that all the regions should be included in the state project plans. When identifying target groups the Government of

India expressed its desire that groups with different socio-economic backgrounds should benefit from the project. For this purpose, specific subsidy and credit schemes were developed for small and marginal farmers, scheduled tribes, and people from hilly areas (GAIC [1982], p. 37).

### 9.3.6 Organisational resources and capabilities

As far as organisational resources and capabilities to be utilised were concerned, a choice was made for central planning (i.e., mainly at the national and, to a lesser extent, at the state level) and for decentralised implementation. Private organisations, non-governmental institutions and government agencies were selected for implementation. It was assumed that they would have sufficient organisational resources and capabilities, skills and personnel to implement the project according to the guidelines issued at the national level.

The selection of leading implementing agencies was the responsibility of the state governments.

### 9.3.7 Summary

The Indian government coordinated all the activities concerning dissemination of biogas technology which had been previously in the hands of regional and local institutions. The coordination was characterised by decision-making on major issues at the national level and implementation at the lower levels. Implementing agencies had limited influence on the overall planning of the project, yet they were expected to fulfil the targets of the national government.

## 9.4 PROCESS OF STRATEGY DEVELOPMENT

### 9.4.1 The role of influential persons and important institutions

Though the process of strategy development was mainly a task of national level institutions and individuals, the first initiatives with respect to development and diffusion were taken at the lower administrative levels. Therefore, it can be maintained that Gandhian organisations and other rural technology institutions, particularly those in Gujarat, have contributed considerably to the initial strategy development by showing the way for mass diffusion of biogas plants.

At a later stage, i.e., after 1960, their role was increasingly taken over by the KVIC, which undertook introduction of biogas technology in a more professional way, supplying more services and covering a broader geographical area than had been possible earlier. This situation prevailed up to the mid-seventies and prepared the ground for a deeper involvement of the Indian government which became interested and supported the efforts of the KVIC, local institutions, and individual beneficiaries (Moulik [1982a]). Finally, in 1981, the Government of India decided to take over entirely the biogas projects of the KVIC. As soon as the government assumed the responsibility for the projects' management, organisational arrangements at the national as well as the state level were made.

Strategy development took place mainly in the Commission for Additional Sources of Energy (CASE) which at that time formed a part of the Ministry of Science and Technology (DST). In 1982, CASE was replaced by the new Department of Non-Conventional Sources of Energy (DNES). In all important matters concerning renewable energy programmes the Advisory Board on Energy, comprising scientists from various state and central government offices, advises the DNES. At the state level, both the Gujarat and Tamil Nadu biogas departments were started in the ministries of the state government.

The influential role of particular individuals as members of government committees or advisory boards should not be underestimated. The late Prime Minister, Mrs. Indira Gandhi, provided strong support for the plan outline. The names of influential members of her cabinet, members of the Planning Commission, and advisors in the field of energy issues who stimulated the idea of a nationwide biogas project should also be added.

### 9.4.2 Use of management tools

Some of the management tools applied in the process of strategy development that were mentioned by Chaudhury [1985] are: needs analysis, generation of alternatives for the option under consideration, and collection of information for decision-making. These three examples are just a few of a long list of strategic management tools that can be utilised to guide the processes of project identification, formulation and appraisal. For this reason, they will not be discussed separately. Instead, the use of strategic management tools will be reviewed in general.

The concepts of project identification, formulation and appraisal are steps in a process in which systematic and scientifically based replies are given to questions that have to be answered before appropriate decisions about the nature of a project can be taken. They form part of what is generally known as 'the project cycle.' For each step of the project cycle, a number of management tools are available.

In the identification phase, the tools of need assessment and problem assessment are used. Project formulation includes technology assessment, sensitivity analysis and cost-benefit analysis. In the process of project appraisal, assessment of the most suitable option and feasibility studies are useful instruments. For all these phases, continuous monitoring and evaluation may be used as aids in the management of projects. A general conclusion about the role of such management tools in the biogas project is that in many cases they are underutilised and even if they are utilised, they are used in a loose and informal way. The impression was that often scientific assessment of facts was considered unnecessary because the authorities involved thought that the answers were already known to them. Quotations of planners and implementors from Blankenberg's study [1985] would illustrate this: 'We felt that biogas would be a good option'; 'Everyone can see that biogas would be a good option'; 'Mr. K. has installed a biogas plant. Why should others not follow his example?'.

The result of this 'impressionistic' approach was that many superficial conclusions about the usefulness and feasibility of biogas plants and the method of implementation were drawn and taken into consideration during the process of planning and implementation. Fortunately, a number of these conclusions proved correct. It is definitely not true that officials are ignorant; often their informal assessments were based on long experience, and consequently, the ideas that were formulated were indeed useful. However, sometimes a more precise analysis should have been made, for example, of the need for biogas plants in different regions or among different socio-economic groups; of the economic cost-benefits of the implementing capabilities of certain organisations; and a realistic assessment of the purchasing power of the poor farmers.

It is difficult to prove that more intensive use of strategic management tools would have resulted in better project designs. However, during the evaluation of the two projects there were many indications that this could have been the case.

### 9.4.3 Summary

In the process of strategy development, local and regional institutions have performed significant preparatory work. The national government, after taking over the project, has greatly benefited from this. Personal interest shown by leading persons of the government bureaucracy in this project has also contributed to its rapid expansion. Tools to guide the process of strategic management were, however, not utilised exhaustively.

## 9.5 PROCESS OF STRATEGY IMPLEMENTATION

### 9.5.1 Nature and role of implementing organisations

The most crucial strategic management decision taken by the state governments of Gujarat and Tamil Nadu was the appointment of certain leading implementing organisations.

In Gujarat, the role of leading organisations was assigned to the Gujarat Agro Industries Corporation (GAIC), though the KVIC was deeply involved as well. The GAIC is a semi-governmental organisation which promotes, develops and establishes enterprises for the manufacture of agricultural machinery, tools, and other inputs. Among its activities is providing managerial and technical guidance to agro-industries. The GAIC can be considered a commercial enterprise. Biogas development was included in its package of activities only a few years ago, but today it forms an important element in the corporation's range of programmes (GAIC [1982], pp. 10, 11).

The KVIC in Gujarat has played the most important role in implementation until recently. Funds for the biogas project are made available by the Government of India through the GAIC. There are a number of local agencies related to the KVIC which are engaged in the installation of biogas plants. The voluntary character of most of these agencies influences the level of professionalism and speed of implementation. Though there are many dedicated workers in the KVIC agencies, they are not always very efficient. The lack of commercial attitude is an important reason for this situation.

There were four major reasons for appointing the GAIC instead of the KVIC: (*a*) The GAIC was expected to be more efficient than the KVIC. (*b*) There were internal problems in the top management of the KVIC in Gujarat. (*c*) The GAIC concentrated its biogas activities in North and West Gujarat and this corresponded with the policy of the state government to put emphasis on these regions. The KVIC was mainly active in South and Central Gujarat. (*d*) The GAIC specialised in the installation of the so-called China-type biogas plant, whereas the KVIC concentrated on its own design. The demand for the first type was growing rapidly (Blankenberg [1983], p. 52).

In Tamil Nadu, the leading agency was the Department of Rural Development and Local Administration of the state government. Again, the KVIC was the second most important agency which was involved in dissemination of biogas technology. The situation in Tamil Nadu was different from that in Gujarat. The history of biogas technology had not been very long and the intensity of activities undertaken in this context was much below that in Gujarat. Tamil Nadu had not witnessed a strong Gandhian movement nor did it have any big commercial enterprise involved in promotion of biogas technology such as the GAIC in Gujarat. For these reasons it was decided that the Department of Rural Development and Local Administration (Directorate of Gobar Gas or Biogas Cell) would play the leading role in the implementation of the project.

The main tasks of this Department are the transfer of funds from the state to the district level; initiation of new activities at lower levels; organisation of training courses; procurement of materials; and coordination of the activities of the monitoring and evaluation committee.

At each district panchayat office, a small cell for promoting biogas plants, which also has the responsibility for implementation at the district level, was introduced. The block panchayats are responsible for extension work and distribution of materials.

Corporate bodies and registered societies play the same role as the block panchayats. Various institutions are active in this way and most of them are related to the KVIC. Annual targets to be achieved are allocated to the block panchayats and KVIC agencies. The target for the former is much higher than that for the latter (Blankenberg [1983], p. 20).

The main reason for the decision to rely heavily on the block administration and less on the KVIC is that biogas cells are represented in all the blocks so that a complete network is available for implementation. The KVIC agencies are not represented in all blocks hence they operate on a limited scale only. However, block officers are responsible for a variety of activities apart from their duties for the biogas project, whereas supervisors of the KVIC agencies are almost exclusively involved in the promotion of biogas technology. The result is that much more work is done by the KVIC supervisors as compared to block supervisors. There is also a remarkable difference in the approach of the block officers and the KVIC supervisors to extension work. The block officers tend to remain in their offices, whereas the KVIC supervisors are more village oriented, often visiting even the outskirts of the villages and beyond. Apart from their greater village orientation, the KVIC supervisors are also better skilled than the block officers. There is no doubt that most of the block officers are not suitable for their job as biogas technology extension workers as a high level of expertise is required.

### 9.5.2 Intra- and inter-organisation coordination

The selection of different organisations also implies that certain assumptions about their coordination and cooperation have been made. In this respect the projects in Gujarat and Tamil Nadu function rather smoothly. There are no major problems between the GAIC and the KVIC in Gujarat. Both organisations work in a more or less clearly defined territory. In Tamil Nadu, the coordination and cooperation between the block panchayats and the KVIC agencies is also satisfactory. Though the block panchayat is obviously the more important organisation, the KVIC agencies enjoy complete freedom of operation.

The situation concerning vertical coordination and cooperation in Gujarat and Tamil Nadu, however, is ridden by problems. In

Gujarat, the relations between the KVIC headquarters in Ahmedabad and the lower level agencies have not been very good for the last few years. As a result of these internal problems, the supply of cement, steel, subsidies, and services has been at a standstill for a long period. Differences in opinion have also created tensions between agencies at the local level leading to an atmosphere in which it is difficult to achieve the set targets. Within the GAIC, such tensions have not occurred; in fact, the corporation has benefited from the internal problems of the KVIC and its agencies.

The project in Tamil Nadu presents another picture. Here, coordination and cooperation problems occurred mainly between the Department of Rural Development and Local Administration on the one hand, and the Directorate of Gobar Gas, regional government institutions and the KVIC on the other. The Department has not shown much interest in the biogas project, yet the Directorate of Gobar Gas has been placed under its close control. Under such circumstances the small biogas cell has had almost no chance of effective operation. Due to this situation problems have arisen between the Directorate of Gobar Gas and the lower levels.

### 9.5.3 Staffing position of implementing organisations

Some remarks should be made about the staffing position of the implementing organisations. In Gujarat, this aspect has been rather well organised: both the GAIC and the KVIC have sufficient and skilled personnel. However, in Tamil Nadu it was often observed that the biogas cells at the district level were understaffed. Frequent transfer of key persons in the Directorate of Gobar Gas had a negative effect on the speed and continuity of work. The lack of skill among the block officers has already been mentioned as a problem, the KVIC was also restricted from reaching its targets due to staff and other capacity limitations.

### 9.5.4 Monitoring and evaluation

The monitoring and evaluation activities in Gujarat were more effective than those in Tamil Nadu. In Gujarat, a powerful state level monitoring committee regularly reviewed the progress of the scheme and the results were discussed with government officials.

One could even say that, thanks to the input of some important members of this committee, the experiences gained in Gujarat played a significant role in planning at the national level.

Within the GAIC, an evaluation survey was undertaken to test the functioning of existing and newly installed biogas plants (GAIC [1982], p. 20). According to the procedures of the government of Tamil Nadu, quarterly and monthly statistical reports as well as quarterly written reports have to be sent to the Department in Madras, but in reality this does not happen. Instead, only monthly reports on the number of applications and budget spent are sent in. A serious analysis of the nature of progress or otherwise has never been made (Blankenberg [1985], p. 127).

### 9.5.5 Summary

In Gujarat, a dynamic commercial enterprise was selected as the leading implementing organisation and this decision proved to be a wise one. In Tamil Nadu, the responsibility was assigned to a department of the state government where a number of problems occurred. The experiences of both projects show that a lack of interest and cooperation between various organisations involved in the project has serious consequences for the progress of activities. The availability of sufficient and skilled staff in the implementing organisations proved a necessary precondition for the achievement of the targets. Finally, serious monitoring and evaluation of problems contributed to improvements in the set-up of the project in Gujarat, whereas in Tamil Nadu these management tools were underutilised.

## 9.6 SOME MAJOR CONCLUSIONS

An evaluation of the performance of the biogas projects shows that the results are not at all poor. One could easily argue that, since the performance is satisfactory, the management must be good. Yet, on the basis of the review of strategic management just presented, it is clear that is not the case. Instead of a systematic identification of favourable and unfavourable management factors on the basis of project performance, it is more significant to discuss the various approaches unrelated to performance. Some major conclusions will be discussed. A choice is made from a number of significant aspects of strategic management.

### 9.6.1 Building on existing networks for project implementation

The Indian government gradually became involved in the biogas project. Preparatory work had been done in Gujarat first by local institutions and later by the KVIC. In the strategy development of the government many of its previous experiences have been incorporated and the existing network of local organisations has been utilised for implementation. However, the government has not been building exclusively on the existing network; whenever required, changes and improvements have been introduced. In Tamil Nadu, the government was not able to build on the existing expertise. The KVIC institutions were involved in implementation but it was realised that the KVIC alone would not be able to install 35,000 biogas plants. For this reason state government institutions were required to play the role that was assigned to the GAIC in Gujarat.

It can be concluded that willingness to learn from the past and building on existing organisational networks and physical infrastructures are some of the positive characteristics of the strategic management approach of the Indian government in this project.

### 9.6.2 The target group approach

Apart from states and regions, the project focused on the distribution of biogas plants among all the socio-economic classes in the rural areas. One of the assumptions underlying the strategy was that even very poor farmers could be reached, and in the formulation of strategy and implementation, certain arrangements were made to achieve the target. These measures, however, were insufficient; almost no poor people have been reached by the project. It must therefore be concluded that the assumptions proved incorrect and that the development and implementation of schemes in this respect formed a weak element in the management of the project.

### 9.6.3 Central planning versus decentralised implementation

Strategic management of the biogas project is characterised by planning at the national and state levels, not only leaving few opportunities for lower level institutions to participate in strategy development and decision-making, but also saddling them with a rather detailed plan of action to be implemented. Though the various regions have different needs and priorities, organisational structures and physical infrastructures, the same plan had to be carried out under all circumstances.

In the initial planning stages of the project, the central government built on past general experiences thereby not allowing sufficient room for local institutions to make useful suggestions or to carry out the work in their own way. Target orientation and strict guidelines concerning the supply of subsidies and bank loans, the supply of materials and the fixed system of reporting are a few examples of the lack of flexibility. This has created problems from time to time, the difficulties faced by the blocks and districts in Tamil Nadu being the most obvious example. It has also resulted in irritation, loss of motivation, and delay in implementation. In Gujarat, where implementing agencies have long experience with biogas technology, dominance by the central authorities and an inability to exercise responsibility were probably felt more strongly than in Tamil Nadu which has limited experience. Central planning in this project indicates weak strategic management.

### 9.6.4 Organisational set-up at the national level

For a project that relies so heavily on national level planning, it is very important that the organisation at the top is perfect. In this respect, the biogas project is sound. Strategy formulation and development as well as planning for implementation were in the hands of a number of personally involved, powerful individuals and officers at the highest level. Representatives of the cabinet, the Planning Commission, and various departments took the major decisions assisted by a capable Advisory Board on Energy Technology (subcommittee on biogas). Technical assistance was also given by various research institutions. With the proviso that the regional and local levels should have been more involved in planning, it may be concluded that the development of a heavy, capable and dedicated organisational set-up at the national level is a strong aspect in strategic management of this project.

### 9.6.5 Packages of facilities

In each project, a careful assessment should be made of all material and organisational facilities required, and action should be taken to provide these facilities. In the case of the biogas project, the following facilities were made available: the organisational set-up, the technology, a maintenance system, skilled personnel, supply of materials, training facilities, a subsidy and credit scheme, an extension service, financial support to state governments, implementing organisations, monitoring and evaluation committees. Though the distribution of facilities and their quality has come in for criticism, the package offered is a complete one. From this perspective, the project can be considered to have a strong strategic management.

### 9.6.6 Use of management tools

The use of management tools as instruments to facilitate the process of decision-making and to diffuse any problems in implementation have been already. discussed. It was concluded that intensive utilisation of such instruments, leading to a more accurate assessment of facts and utilisation of this information, will have a positive impact on the performance of projects. Some examples have been cited of situations where a more serious utilisation of such tools could have resulted in better planning and implementation. Reference was made to need assessment, economic cost-benefit analysis, assessment of capabilities of implementing organisations, and assessment of resource constraints among particular target groups. Hence, the conclusion can be drawn that optimal utilisation of available management tools is a weak aspect of strategic management in the biogas project.

### 9.6.7 Role of leading implementing agencies

The selection of the GAIC as an implementing agency in Gujarat proved a good decision for the following reasons:

1. It is a commercial organisation, which had positive effects on the number and quality of biogas plants installed.
2. It has an appropriate organisational set-up, with offices in the various concentration areas of the biogas project.
3. The regional offices have a multi-purpose function. Many farmers, visiting the offices for a particular business purpose, become automatically acquainted with the activities of the GAIC in the field of biogas technology. Moreover, various activities take place in the villages providing frequent contact with the target groups. The GAIC offers a total package of facilities for the farmers who show an interest in a biogas plant and the corporation works fast. It conducts training programmes for the staff, engages in research and development activities and emphasises the introduction of the popular China-type biogas plant.

The selection of the Department of Rural Development and Local Administration in Tamil Nadu, though perhaps necessary in the absence of serious alternatives, was not so successful. There are several reasons for this:

1. The Department did not show much interest in the project, with the exception of the Directorate of Gobar Gas, where job motivation slackened as a result of the tight control on all the activities and non-cooperation at higher government levels.
2. There was lack of skilled personnel at all levels of the government system and frequent personnel transfers further weakened the position of the Directorate of Gobar Gas.
3. The Department was unable to supply subsidies and construction materials on time.
4. The monitoring and evaluation system did not function properly.
5. Block offices were supposed to perform the tasks at the lowest level, but they had too many other tasks to pay sufficient attention to the biogas project.
6. Communication, coordination and cooperation were weak.

Aspects such as attitude, dynamism, speed, organisational

structure and capabilities, field orientation, staffing, communication, coordination, cooperation, and feedback should play an important role in the decision of project planners to select a particular organisation as the leading implementing agency. However, as can be seen from the foregoing review, this obvious conclusion is not in all cases drawn by policy-makers in India.

## NOTES

1. Personal communication from Inderjit Singh Anand, Deputy Chief of the Rural Electrification Corporation, New Delhi, June 1988.

# 10

# BIOGAS TECHNOLOGY: IS IT APPROPRIATE?[1]

**Kanta Ahuja, M.S. Rathore and Deepak Gianchandani**

A general feature of small-scale decentralised technologies is that they tend to be highly location specific and require far more attention to locationally variable factors. The appropriateness of any technology has to be assessed carefully in each sub-regional context. The appropriateness of a strategy can be evaluated from several different points of view which do not correspond in every situation. A strategy may be regarded as appropriate from the long range view of the planner but inappropriate for those who have to implement it. And, even if it is appropriate for planners and implementors it may turn out to be inappropriate for those who are at the receiving end of the technology. Given these assumptions, the nature of the problems faced at the village level can be categorised into three: (*a*) problems that originated in the planning stage, (*b*) difficulties due to the implementation process, and (*c*) operational problems at the user end.

## 10.1 PLANNING: OBJECTIVES AND FOCUS

From the planning perspective, the aspects that are particularly crucial are those relating to the technology itself, those pertaining to the specific focus for promoting the technology, and those concerning its viability. A judgement on the purely technical aspects of the design being promoted in India, is beyond the scope of the present study. However, one remark must be made about the basic design currently being used. There is scope for altering and improving the design based on further R & D in order to suit local environmental conditions, including differences in available material to feed (feed lot) the biogas plant. The appropriateness of the biogas technology is established in terms of its simplicity, low cost and related advantages such as the use of bio-fertiliser, waste recycling, public health, environmental management and pollution control. Despite individual benefits, there is still the question of specifying the precise focus of the strategy. This will determine the approach to motivate the users, as well as the assessment of costs and benefits.

The three principal outputs, or outcomes, of the technology are (*a*) energy, (*b*) fertiliser, and (*c*) improved hygiene. The energy gain is largely from the long range perspective of the planner. From the point of view of the user, the energy gain depends upon whether or not biological and farm wastes that were not normally used for burning, become useful through the biogasification process. If only farmyard manure is used as feed lot material, there is very little immediately perceptible energy gain. There is merely a substitution of one form of usage (dung-cake) for another form (gas). The gain is mostly in terms of convenience. However, the perceptions of the user are far more important here than those of the planner.

The technology is being promoted as part of an alternative energy programme. However, the output that is perceived to be most valuable by the user is slurry which is limited to the double cropped irrigated areas. Technology should not merely be transferred; it has to be adopted. This happens when users begin to identify themselves as a part of the technology and the benefits are recognised. If this is so, it follows that both plans to motivate the users as well as the extension methodology need to be adjusted to project the advantages of fermented organic wastes for the fertilisation of land. It would also mean that the implementation will have to be linked to agricultural extension networks rather than be a separate part of the rural development network.

In this context, it is interesting to note some of the aspects of the programme implemented in China where biogas was developed primarily to control parasitic defecation in open fields. The hygiene and environmental aspects were considered as important, if not more important, as gas generation (Rajeswaran [1983], p. 31). At the planning and implementation stages for popularising biogas technology, the participation of prospective users was ensured in all aspects of design, construction and maintenance. This was considered an essential input in training, skill formation and human resources development. Labour costs, therefore, were seen not as costs but as an investment in human capital formation. The focus of the approach has been on motivation and training at the receiving end, and on research and development of suitable types of plants for different environmental conditions at the delivery end. As a result of this approach, 7,000,000 plants installed by 1978 could be supported by as few as 200,000 technicians, which is a ratio of one technician to 350 plants. The effort is supported at the commune level by each production team with at least one member who is well trained in all aspects of the technology. This person receives continuous training and support through participation in discussions and workshops in which experience is shared and problems sorted out.

In India, the technician to biogas plant ratio is estimated to be only 1:60, and the technological support for the programme at the supplier end and at the user end is minimal. In short, inadequate attention to detail and a lack of correspondence between the anticipated benefits and the realised benefits result in a motivational strategy which faiis to evoke a response from the users.

## 10.2 VIABILITY

The viability of the technology is another important aspect. Once again, the planners' macro-assessment may not be correct for every local situation. The technology will be most viable when the alternatives to the input have a high opportunity cost. If the output is cooking fuel, the relevant question for determining viability will be the specific cooking fuel being replaced in a given situation. Is there a fixed limit to the amount of fuel available? Are there problems due to insufficient cash or inconvenience due to transport? These factors are likely to vary from location to location. Viability also depends on whether or not the essential inputs have a low opportunity cost. The inputs in this case are feed lot, water, and availability and willingness of labour to operate the plant. Operational efficiency, in turn, depends on siting, continuous input, plant size, temperature, water availability, avoidance of scum formation and technical expertise in construction, maintenance and servicing.

Within this general context it becomes possible to understand why the plants may have run into difficulties in Rajasthan. The opportunity cost of alternative fuel, which in most parts of Rajasthan is farm waste and dung and not fuelwood, has till recently been relatively low. The other local factors are a low population density, a scattered settlement pattern, and lack of water. Lack of water leads to a low vegetation density so that there is not enough material for biogas digestion. Livestock numbers are large but, in terms of body weight of animals, the livestock is predominated by smaller animals and depends largely on dry feed materials. Stalling of cattle is virtually non-existent. All these factors reduce the availability of feed lot material per plant. The labour input in collection of farm wastes for direct use as fuel has a low opportunity cost, but there is no reduction in this input; it takes a different form. The opportunity cost of inputs in biogas and of any of the alternative fuels is roughly the same. On the other hand, the cost of the plant even with the subsidy is excessive in the context of the total farm enterprise. The estimated cost of even the smaller two cubic metres plant is between Rs. 4,000 to Rs. 6,000. This is a major item in any situation. Any unit costing this much requires justification from the farmer's viewpoint. It is not surprising that one respondent asked the question: 'Why should I not dig a well with this money?' When the farms are dry, adding even two buckets of water in a unit may be seen as non-essential. The installation of a plant is hardly perceived as a felt need or as a priority. Farmers are practical, realistic people. Therefore, unless a technology is viable at the level of the individual it cannot become popular simply because there are macro-advantages.

The technology must be seen from the perspective of the user and not merely from the perspective of the planner. If the latter's assessment has been inaccurate, the ensuing implementation invites mismanagement and a vicious circle sets in, whereby it is difficult to say whether the failure is due to mismanagement or some more essential problem. The incorporation of the perspectives of the users necessitates participatory decision-making. Furthermore, the process of technology transfer is not simply a process of constructing a plant and funding it. The real problem is an extension problem and the technology transfer must be linked to an extension network. If, in the Indian situation, the real advantage lies in a better quality of organic manure obtained from the plant, either in the form of slurry or as a result of slurry mixed with other organic materials, then there is a need that this aspect be fully tested. It must be examined by agricultural scientists and made a part of the agricultural extension system rather than as part of a strategy of saving domestic fuel. When the purpose is to save domestic fuel, a low cost pressure cooker may yield better results. If the gas fuel output is the primary goal of the technology, then it should be popularised in units where the alternative fuel input constitutes a major input either as motive power or for lighting. Dairy and poultry enterprises or mixed farming systems with high density in terms of vegetation and animals would gain more than mono-crop dry farming systems. Such an approach would require more R & D activity to enhance the use of poultry waste in the digester and better use of gas for lighting in the poultry pens.

## 10.3 IMPLEMENTATION

The implementation problems are related to a diverse set of problems. Many of them are specific to particular situations. The general aspects of the implementation process that need strengthening or rethinking are (*a*) technical support for construction and maintenance, and (*b*) the role of the subsidy, its form and the process of disbursement. The plants in Rajasthan are constructed under government supervision, and the beneficiaries play a passive role. Therefore, the large number of operational failures are related to faulty construction, use of sub-standard or inadequate material and failure to provide burners and related fittings. Subsidy provisions create a host of problems including improper selection of beneficiaries. As a motivation device, it helps to reach targets but it generates expectations which cannot be fulfilled. The subsidy is an arrangement for sharing of costs, but it is expected as cash assistance. The users are not made aware of the 'costs' involved. These costs are real costs in terms of the adjustments required in work patterns and living styles. Back-up support for repair and maintenance is virtually non-existent.

## 10.4 USER-END PROBLEMS

The user-end problems are due to the lack of familiarity and knowledge of what to expect and what is required for successful operation of the plant. This can be illustrated by several examples from the field.

In one case, bricks and stones were piled on the drum to prevent it from rising, so that the pressure with which gas comes out is increased. It is a common practice to put urea in the digester to increase gas formation, lime is added for the same purpose. This leads to a short-term increase in the quantity of gas obtained but leads to destruction of bacteria and reduces the production of gas in the long run. In another case, a match was lighted near the outlet to find out whether gas had started forming. An explosion and damage to the plant were the result. In one village owners expected gas as heat; the outlet was repeatedly tested to feel if it had become hot.

Seemingly irrational biases sometimes created problems in operating the unit. For example, biogas is still rejected for cooking their own food by many. However, the inadequacy of the amount of gas produced would in any case have led to allocating priorities for the different uses of gas. If gas was available in adequate quantity, it is possible that the bias would have been overcome with time. In many cases this did happen. Thus, the real problem is not the bias out inadequate gas production. The lack of training and familiarisation with the technology and the fact that there is no follow-up after the plant is installed has led to unnecessary closures. Often the pipeline was of sub-standard quality and gas leaked from the pipeline. In one case, the inferior quality pipeline had been wound into a roll and hung on a tree, which blocked the flow of gas. The corrective action required in these cases is very minor and can be taken easily if the beneficiaries are familiar with different parts of the plant. But beneficiaries are told not to alter anything and wait for someone from the block office to look into their problems. When no one arrives, they close the unit and stop using it. The impact of this upon others who then close their plants too creates a negative image and a rejection of the technology.

There is considerable work involved in operating the plant properly. Dung has to be mixed with water and care has to be taken to prevent sand and foreign material from getting into the digester along with the mixture. This mixing is usually done by hand and the operation is unpleasant and initially unfamiliar. When men undertake this task it means more work for them since the collection of dung and making cakes are considered a woman's job. If women have to perform this task, they also find it less acceptable. The work related to the biogas unit adds to their existing household chores, moreover they were not involved in decision-making and training when it was introduced. The increased workload is further compounded by the inadequate amount of gas obtained. The quantity is usually less than their requirement and the fuel does not allow 'batis' and 'rotis' to be cooked in the familiar way. Furthermore, because the supply of gas is beyond their control and they do not know when it will be exhausted, they still have to maintain a stock of dung-cakes or fuelwood.

Gobar (dung) is used in the villages for many purposes such as fuel and manure and for plastering the walls and floors of huts. The ashes are used for cleaning utensils and as an insecticide in the fields. The traditional *chulha* also provides some warmth during winter. Compared with this, the biogas unit produces only two direct outputs—fuel and slurry. Fuel availability is often inadequate and slurry has to be mixed with organic manure. The traditional sources of fuel, like dung-cakes, waste wood and twigs are obtained by wome· and children. From the household's perspective the labour involved in this does not entail any cost. The perceptions of costs and benefits are such that the costs of traditional technology, such as the gathering of dung and twigs are not regarded as costs and the benefits of biogas, such as convenience in cooking, are not perceived as benefits.

The livestock related problems are no less important. The biogas unit entails changes in feeding practices; animals have to be stalled. This requires, apart from some adjustment in the work related to feeding and grazing, changes in the composition of the herd in favour of a larger number of buffaloes. Cows can be stalled for only limited periods because they need grazing for biological reasons. Buffaloes are more suited for stall feeding. In order to obtain a regular supply of milk, three or more buffaloes are required as they miss calving more often than cows. A two to three cubic metre plant, requiring two buffaloes, is recommended for small beneficiaries. Two buffaloes, however, are not enough for a regular supply of milk. If a beneficiary is not able to maintain the required size of the herd over a long period of time, he probably has to sell the two buffaloes with which he started. Thus, the biogas unit becomes non-viable, though it may have been viable earlier.

The availability of dung depends on the number of cattle and on the grazing practices. This is recognised in the literature on biogas plants. What is not appreciated is that there are other factors such as shortage of green fodder and the migration of cattle over short or long distances during the summer months. This reduces the availability of dung during the season when the plant is likely to operate most efficiently.

Whenever a farmer appreciates the new technology he also tends to show innovativeness and resourcefulness. For instance, a locally fabricated churner was used to mix dung with water. It reduced the unpleasantness of work, speeded up the process, rendering men more willing to do this work. In one village, the burner had been improvised by making holes in a tin top and attaching the pipe to the bottom of the tin. Another beneficiary had substituted the existing outlet by non-rusting brass outlets in order to remove the water. In another village, the gas plant was on the ground floor and the kitchen was on the first floor. A pipe had been successfully fixed to provide gas.

Villagers are equally innovative in finding alternative uses for plants that are not used, for example, as a place to store fodder and grain, or water. In one village it was even used as living quarters, which were very cosy in winter and cool in summer.

## NOTES

1. This chapter is based on a field study for evaluating the progress of the National Biogas Programme in Rajasthan. The study was undertaken on behalf of the Centre for Studies in Decentralised Industries, Bombay and was sponsored by the Department of Non-Conventional Energy, Government of India. It covered six districts of Rajasthan in which the biogas programme was implemented between 1980–81 and 1982–83. While general applicability of specific conclusions is not claimed, the conclusions are presented in the hope that they will generate insights into the general problems of technology transfer and technology adoption in rural areas.

# 11

# WOMEN AND COOKING TECHNOLOGY: THE VICISSITUDES OF IMPROVED STOVE PROJECTS IN RURAL GUJARAT

**Loes Schenk-Sandbergen**

## 11.1 PERSPECTIVE AND METHODOLOGY

This study presents data from an investigation of the role of women, especially poor rural women, in projects and programmes which distribute improved (smokeless) *chulhas* (ovens). The main objective was to provide insight into the conditions under which women of poor households can be reached with more success by such programmes. This research aim is based on the assumption that improved chulhas enhance the living conditions of women of poor households by saving time because cooking can be done faster on the chulha, saving firewood, causing less eye irritation because of less smoke or soot, leaving the kitchen cleaner, etc. The success of a chulha project is assessed from the users' point of view, by examining the experiences, opinions and behaviour of the involved women. When they accept the new technology, using it regularly over a longer time span and considei that it provides some relief from their household chores I regard a project as successful.

### 11.1.1 Perspective

The National Project on Demonstration of Improved Chulhas (NPDIC), designed by the Department of Non-Conventional Energy Sources, was launched in 1983. It claimed that the introduction of chulha stoves would mitigate the hardness of women's lives, improve health status and, above all, save fuel, especially wood fuel. The programme was announced as a programme 'for women by women' and aimed at 'helping to solve the cooking crisis'. Madhu Sarin [1986] cites an estimate that the total number of chulhas built by March 1986 was apparently 2,000,000, with the expectation that 16,000,000 tons of fuelwood would be saved. She argues that the last figure is far too optimistic. A programme on such a large scale would require a very effective approach to be successfully implemented which depends, according to Bina Agarwal [1986a, p.70], on the technical, economic and social characteristics of the innovation. In the case of improved chulhas these characteristics include:

1. 'Service' technology.
2. The gender determinant.
3. The caste/class dimension.

#### A 'service' technology

The 'new' cooking energy technology differs substantially in character from the 'Green Revolution' or 'production' technologies which have transformed agriculture into an agro-business. These technologies have spread on the basis of the market mechanism as profit-making was involved. The improved chulha technology is different as it is a 'reproduction' or 'supporting' technology. In the Indian setting it is not likely that the marketing of chulhas on a commercial basis will be possible. Therefore a 'guided introduction' or non-profit management of the introduction of the stoves by government agencies, voluntary organisations and, in some cases private companies will be necessary. Although 'production' and 'reproduction' technologies require different approaches of implementation, they are closely linked in the sense that the 'reproduction' technology is subordinate to the 'production' technology. The capitalist mode of production demands an improvement in the near future of 'production' technologies such as harvesting equipment, threshing machines, tractors and irrigation facilities—which all need energy. Alternatives in energy sources, such as non-conventional energy, are being increasingly proposed as a means of continuing the process of agricultural transformation. It is in this context that a whole range of new rural energy technologies has been planned and designed and large programmes have been implemented to disseminate these technologies at the village level, for example, through 'integrated rural energy centres'. At this juncture I would like to point out that the notion of the urgent need to save wood fuel (trees and forests) as a precious energy source by introducing a massive wood saving chulha programme, did not arise in the first place out of a concern to alleviate the drudgery of poor women or a concern for vanishing forests, but out of concern for the needs of capitalist and commercial forces. However, the successful implementation of alternative sources of energy should begin with the problem of women of rural households facing an increasing shortage of wood fuel.

**The gender determinant**

'Women ' and 'technology', as related subjects, have recently received considerable attention in the literature and in debates on alternative strategies in development. There is an emphasis on analysing the needs of women and the problems they face, especially women of poor households, in order to understand the process of acceptance, utilisation and dissemination of 'new' technologies. Women have traditionally played a strong role in technology decisions. Translation of these traditional responsibilities into roles in recent projects of 'new' technologies has certainly been the exception, but could be a key item for 'strategic' management concepts. It is virtually universally true that poor women are exclusively involved in the managing of cooking energy. Due to the division of labour and the structuring of gender relations it is their responsibility to gather cooking fuel, cook food, heat water and perform many other tasks related with energy. Recently Bina Agarwal [1986a] documented the serious problems poor women face because of the wood fuel crisis in the Third World. Their burden is increasing because of deforestation and the consequent decrease in the absolute availability of wood for all purposes, which is related to the over-exploitation of waste and common lands for various purposes. Furthermore, traditional cooking methods expose women and children to serious health hazards. Therefore, it is a precondition that an improved chulha project has the support and cooperation of women and children of poor households (Bina Agarwal [1986a]).

**The caste/class determinant**

During field visits in Gujarat it was clear that the improved chulhas are unsuitable for extremely poor families and other particular groups like nomads, who can only afford to cook a one item meal in one pot. As they travel frequently the fixed chulha with a chimney is obviously not useful. Moreover, they cook mostly outside the hut. For rich households the 'new' chulha programme also has little relevance. In Kheda district, Patidar houses belonging to rich farmers have many modern cooking facilities: kerosene stoves of all varieties, solar cookers, pressure cookers, gas cylinders, gas stoves and even private biogas plants providing cooking gas and light. The general notion often noted in women studies that only men benefit from modern technology and that women are forgotten has to be modified. Much more emphasis should be given to class/caste differentiations among women.

The large majority of poor women cannot buy modern cooking equipment and use the age-old mud stoves thereby facing all the drudgery and inconvenience known by generations before them. Hence, it can be expected that they will be the caste/class potentially interested in projects to improve their cooking fuel situation. However, there was a case of a Patidar woman, who had the entire range of modern cooking facilities in her kitchen as mentioned earlier, who continued to cook on her somewhat improved traditional mud stove, because, as she explained: 'it is the cheapest way'. She could get wood and agricultural wastes free from 'her land' and use it as cooking fuel.

In the foregoing, three conditions for the successful implementation of chulhas were discussed. The gender dimension will be dealt with in more detail.

### 11.1.2 The problem

It is one thing to design a programme 'for women by women' but it is another to implement it. In the majority of cases, the type of technology is determined by men, who also constitute the management of the project which introduces the 'new' technology. These men are largely from urban areas with a high educational level and high status. In the implementing organisation, men are the managers as well as field and extension workers. Usually women are hardly involved in these areas because of the constraints related to their position in society. Hence, two important sets of factors need to be borne in mind:

— Women are rarely the initiators, not having access to 'new' technological knowledge, nor to the resources and opportunities necessary to use this knowledge. The result is that technological projects are designed without regard to women's real requirements and priorities.
— Women face the problem of sex segregation which has its roots in attitudes of role differentiation and in the concept of female modesty, which every girl learns.

According to Ursula Sharma [1985, p. 270], the segregation involves

> . . . her whole demeanour in the presence of men. Thus, in any neutral public space in which men and women are likely to interact—such as the area round a bus stop, the verandas of village grocery stores, the open space near wells or the village streets—it is the business of the women to protect themselves from unnecessary contacts with men.

The rules of segregation are asymmetrical, as men are obliged to a much lesser extent to withdraw from interaction with women. This situation is reinforced by women themselves, as a woman who does not maintain a discrete distance from men in public will be severely censured by other women (Sharma [1985], p.270). The idea is influential that, for a woman, activities or work which involve intensive and public contact with men, would be immodest and 'unnecessary'. It is not that women are considered to be incapable or incompetent but the concept of modesty ensures that women are excluded.

From the above it is obvious that male technicians, field and extension workers cannot just enter the kitchen of a woman and construct an improved chulha; neither can they call upon the village women to get their opinion about the best location of the water tap and then return to their office; whether they would be prepared to do this is another question as it has implications for their status. Thus, the problem of a sharp distinction between men's and women's spheres restricts the interaction and communication between men implementors and women users, all the more when men implementors deal with sectors which are directly related to women's domains and responsibilities.

Furthermore, there is the problem that women are often not allowed by their parents or husband to be employed in organisations functioning at the village level. Sex linked roles associate women primarily with the home and men with the outside world. These problems greatly hinder organisations implementing a chulha programme 'for women by women'.

### 11.1.3 Methodology

When collecting data I focused on five projects in Ahmedabad, Kheda, Baroda and Panch Mahal districts. The choice of the location of these projects was based on one main variable:

1. The incorporation and involvement of women in different functions in the implementing organisation.

I have related this variable to two background variables giving the contextual setting:

2. The general level of socio-economic development in the district.
3. The nature of the implementing organisation.

On the basis of the identification visit in Gujarat in 1984, I selected three projects in which women were involved in several functions in the implementing organisation. I differentiated the following functions: director (leader), initiator, propagator, designer, demonstrator, constructor and maintenance worker. The selection of these projects implies an alternative approach because, unlike other projects in the past, they are not fully male dominated and male directed. For reasons of comparison I selected two other projects in which only men were employed in the same functions.

To have more insight into the relation between the level of socio-economic development of a certain area and the involvement and incorporation of women, the following districts were selected: Kheda as the most developed area, Baroda and Ahmedabad as developed areas, and Panch Mahal as a less developed area. In general, it can be expected that local women in less developed areas face more problems in finding employment because of transport difficulties and fewer educational facilities. On the other hand, I was aware that in the less developed area many villagers are of tribal or semi-tribal origin. It is known that in such groups women have relatively more freedom of movement than Hindu women in rural areas.

What is the relation between the nature of the implementing organisation and the number of women working as volunteers or employees in the organisation. Three types of implementing organisations were selected.

— Government: The Forestry Department is an important implementing organisation which executes the national chulha programme.

— Private company: The government gives tax concessions to companies which establish voluntary organisations for rural development. Many companies have responded to this concession for a variety of reasons, from idealistic to purely financial and commercial interests.

— Non-government Organisations (NGO) or Voluntary Organisations (VO): I came across a wide variety of NGOs and VOs which intended, among other activities, to improve the cooking fuel situation. Smokeless chulha and solar cooker demonstrations were, hence, very widespread. The organisations differed in scale, scope, objective, approach, financing sources, ideology and structure. In the context of this study I selected one VO which functioned as an intermediary organisation and two NGOs which operated directly at the grassroots level. The difference between an NGO and a VO in this context is that the former is foreign funded and the latter is not.

Of the five projects I studied three were implemented by NGOs/VOs, one was implemented by the Forestry Department and one by a private company.

A classification of the projects according to the above variables is as follows:

| *Location* | *Female Involvement* | *Socio-economic Development* | *Type of Organisation* |
|---|---|---|---|
| Kheda | Yes | High | NGO |
| Baroda | Yes | Medium | Private Company |
| Ahmedabad | No | Medium | VO |
| Panch Mahal I | Yes | Low | NGO |
| Panch Mahal II | No | Low | Government |

To assess the success or failure of the five projects, field visits were made in 1985 to each project area. During these visits I collected socio-economic data, observed the situation in the kitchen and particularly noted the condition of the chulha. As the success of the chulha project is assessed from the users' point of view, many women (and men) were interviewed about their experiences and opinions regarding old and new stoves, particularly the advantages and disadvantages. When they accept the new technology, using it regularly over a longer time span, and consider it as providing relief from their drudgery in cooking, the project is regarded as successful. I also had discussions with groups of women from poor households in the villages visited.

In three locations women were selected on the basis of the socio-economic condition of the household (the type of house, compound, etc.) and whether there was, or had been a smokeless or an improved chulha. In one location, more structured interviews were possible because we were able to stay in the village for a week.

It is obvious that, besides the role of women, there are many other variables that have to be taken into consideration in obtaining an insight into the conditions for the success or failure of a project. These variables will be discussed later. Nevertheless, the emphasis will be on the proposition that the success of chulha projects will increase when local rural women are involved and included in projects where they perform different functions. A brief outline of the background and problems of the cooking fuel situation in

Gujarat and the advantages and disadvantages of improved stoves will be presented first.

## 11.2 FIREWOOD SCARCITY AND IMPROVED CHULHAS

### 11.2.1 Cooking fuel and women in Gujarat

Landless labourers' families collect most of their own fuel free, whereas big farmers generate their own fuel from crop residues or purchase it. Women of landless labourer families spend, according to a survey, three to five hours a day collecting cooking fuel (GEDA [undated]). The cooking time varies from 3.3 to 3.6 hours per household per day. A rough estimation is that a family of five members needs at least 20 kg of wood fuel in four days (Foley and Moss [1985], pp. 59-60). Cooking fuel used per capita per year varies from 373 to 420 kg, that for water heating and winter 'warm up' accounts for another 10 per cent.

My observations on the management of cooking fuel illustrate the situation in a village in Baroda district.

> Wood, branches, twigs and crop residues (tur and cotton stalks) and very few cowdung-cakes are used as fuel. Primarily women and girls collect wood fuel in the jungle area at a distance of two to three km from the village. Although the distance is not far, the area is hilly and the soil is scattered with uneven stones.
>
> It is common for women and girls from one house or neighbouring houses to go in groups, once in three, four days to the jungle area, accompanied by men or boys. But I also saw two women, or even one single man fetching firewood. It seemed that the sexual division of labour concerning collecting wood fuel was not so strict. When I met a man with a bundle of firewood on his head, he told me that in his household the collection depends on the workload and place of work of the household members. Coming from his fields he is near the jungle area, thus, he brings the firewood home.
>
> There was a clear structure in the system of collecting wood fuel regarding the types and varieties collected. From one group some women carried a bundle of only long and thin twigs, another woman carried only thick and short pieces of wood in a basket on her head, and the men carried heavy trunks together

with the axe which they had clearly cut off branches of trees. This collecting of different varieties of fuel can be related to the food that has to be prepared. 'Chapatis' require a fierce heat and trunks are the best for that cooking purpose. Boiling of rice or jowar requires a low fire which can be fed with wood twigs or tur or cotton stalks. These differences in wood fuel needs explain why women from landless labourer families replied to our question: 'How many times a week you collect fuel?', with 'that depends on the food to be cooked'.

Households with cattle and land use also some cakes made of cowdung 'chanaas', but in general the cowdung is used as manure. These cakes are made exclusively by women, at the cost of their labour and time. The children of poor herding families have the right to pick up the droppings of the cattle they take to the panchayat wasteland. They bring it home or exchange it as manure for some food items. In winter time the women make a stock of wood fuel before the busy agricultural season will start, especially the monsoon time. As wood is difficult to light during the monsoon they have to use some cowdung-cakes but by experience they know that this fuel gives a lot of smoke and therefore they try to reduce that disadvantage and add some thin twigs to have better cooking efficiency (Schenk-Sandbergen [1985]).

In a study on women's drudgery in firewood collection in Gujarat (Nabrahman and Sambrani [1983]), four villages with varying proximity to forests and of different ethnic background were investigated. The study indicates that among the poorer tribal households, difficulties in collecting firewood and other cooking fuel are on the increase, thus increasing women's drudgery. Poor women face a growing scarcity of cooking fuel which is caused by:

1. The present consumption of energy (especially the needs of the cities) leading to deforestation.
2. The commercialisation of wood resulting in widespread deforestation.
3. Forest conservation depriving poor tribal women the customary right to gather cooking fuel free.

4. Alternative uses for cowdung and stalks as valuable commercial materials for other sectors of energy supply and development projects also exacerbate the scarcity.

In Gujarat, there is a high correlation between land ownership and access to firewood and other biomass fuels. In villages in Kheda district the panchayat (village council) exercised control over the trees growing on the communal wasteland. Poor people were forced to buy firewood from the panchayat but often did not have enough money and could only light their cooking fire once a day by burning shrubs, small twigs, etc. gathered at the roadside. The poorer the woman, the more restricted is her access to cooking fuel, the longer her working hours, and the longer the distances she has to walk carrying heavy loads every day.

All over India there is a rapidly increasing demand for wood as a raw material for industry, house construction, packaging etc. This has resulted in widespread deforestation. Consequently, the village housewife now has to trudge long distances to collect her daily requirement of fuel for cooking.

Nabrahman and Sambrani [1983] observe that the necessary and logical policy of the government on forest conservation has ironically led to increasing drudgery for rural women. Even those living at the edge of the forest, from where they used to collect free firewood in the form of decayed branches and tree trunks, have found that their access to such supplies is now blocked. Therefore, rural women have to work harder to achieve the same results. In Baroda district, at the edge of the forest hills of Panch Mahal, the situation differed from the one described by Nabrahman and Sambrani [1983]. In the village it was common knowledge that the forest authorities were liberal and allowed the collection of firewood without asking for any money or anything else as long as people carried the wood load on their heads. They would be fined, if they transported the wood in a cart or a lorry.

Poor rural women face an increasing scarcity of cooking fuel as a result of 'new' technologies in other sectors of energy supply. For example, to supplement the fuel available in the rural areas, utilisation of animal wastes simultaneously for both manure and fuel purposes through biogas plants may be considered laudable. At the same time, however, this realisation of better use of what has hitherto been considered waste, has caused deprivation to those dependent on its free supply, namely, the poor households. Similarly, animal husbandry in rural areas may be considered extremely desirable from the point of view of augmenting meagre incomes. As a result of the necessity to feed animals, however, crop residue is no longer easily available for fuel purposes (Nabrahman and Sambrani [1983], p. 38).

### 11.2.2 Impact of cooking fuel scarcity

Cooking energy is an important factor determining the nutritional level of the family's food intake. Scarcity of fuel forces women to reduce the energy used in cooking with the resultant nutrient losses and greater risk of infection and illness. Undercooking of oil even leads to the danger of food poisoning. Dietary patterns and nutritional status are considerably affected by access to cooking energy. Fuel shortage and its impact on nutrition is, in practice, more serious for women than for men of the same social category as men are traditionally served first.

### 11.2.3 The advantages of improved chulha projects

Poor women have typically one 'chulha' which has extremely low cooking efficiency. It is estimated that the heat efficiency of this type of stove is 10 per cent, thereby implying that some 90 per cent of the heat of the wood that is burned is wasted (Chauhan [1979], p. 9).

What can be the benefit for poor rural women of adopting an improved or smokeless chulha? The literature reveals the following positive aspects;

1. Reduction of fuel consumption.
2. Reduction of smoke.
3. Decrease in labour time spent on cleaning clothes, walls, etc.
4. Reduction in scrubbing blackened utensils.
5. Decrease in cooking time.
6. Improvement of women's status.

Madhu Sarin describes the benefits of the improved chulha:

> It can reduce the time and effort spent on cooking. More time can become available for better child care, productive work and rest . . . As young girls often do the cooking, it can mean a healthier new generation of women. Most of all, as the technology mainly benefits women, the improved chulha can help overworked and exploited rural women get a new consciousness. It can make them want to improve their lives and status in society (Sarin [1984a], p. 6).

However, there are also advantages of traditional cooking on open fires which are partly lost with the introduction of smokeless chulhas:

1. An open fire can burn fuel of any size, shape or type and where cutting tools are primitive it may be the only practical way of using large pieces of wood.
2. It can be located anywhere to take advantage of shelter or shade.
3. It can be built free of cost with local materials (clay).
4. Open fires give more heat and light than smokeless chulhas.
5. Smoke has its use in preserving seeds for food crops and it can destroy termites and other insects in thatched roofs and walls.

It should be realised that the advantages of improved chulhas are dependent on sound construction and correct use. In practice, however, many construction faults arise due to ineffective or lack of training of users, or inappropriate materials and parts such as a chimney with too large a diameter.

The consequences of lack of user training may be that there is an increase in fuel consumption and that women have to collect more fuel instead of less. Madhu Sarin writes:

> The biggest tragedy is that the worst sufferers are the poorest people. The better-off families do not mind increased fuel consumption as they get enough fuel from their land and animals. They are generally pleased with the benefit of smoke removal. But the poorer people have to scrounge around for fuel. They can ill-afford doubled or trebled fuel consumption as it increases their labour. Further, as their houses have thatched

roofs, they are unwilling to enjoy the benefit of smoke removal at the risk of their homes catching fire. The fire hazard *can* be avoided with careful construction and proper users' education (Sarin [1986], p. 1741).

This means that poor women can be disadvantaged in a serious way when they accept, or are pressed to accept, an improved chulha.

## 11.3 CHULHAS AND AN INTEGRATED RURAL ENERGY CENTRE IN BARODA

### 11.3.1 Energy and the village

In Baroda district, in a village of 100 households, 50 smokeless chulhas had been introduced some time ago. However, the developments in this village are exceptional. A whole set of 'new' technologies in every field of daily life has been, or will be introduced as the village has been selected as an 'Integrated Rural Energy Centre' whose energy needs are to be based on renewable, non-conventional sources of energy. These include a gasifier, a biogas plant, community solar cooking, solar cooking, solar steriliser, solar hot water system, etc. This project is accompanied by an intensive programme of rural socio-economic development, which had been launched in 1980, making it a model village of Gujarat.

A range of innovations had been implemented: an overhead tank and a 3.5 KW diesel engine pumpset, while a hectare of trees had been planted. The village panchayat has granted 12 hectares for an afforestation programme. Latrines have been installed. A kitchen-gardening and educational programme for women and girls is being conducted, and balwadi (kindergarten) and food programme have been introduced recently.

It is a very interesting village because of the organisational structure of the rural energy centre. Designing, developing, installing and making the technologies workable will be the responsibility of the energy division of a private company. The participation of the people and all services will be organised by the agricultural division of this company. Furthermore, there are two voluntary agencies linked with this private company which also work in the field of health, hygiene, education, social development, agriculture and animal husbandry.

Contrary to the usual approach of introducing piecemeal programmes adopted in the absence of a broader perspective, in this village an effort is being made to implement technological innovations, together with social and economic activities which are intended to strengthen the weakest sections in the village. However, it should be mentioned that, in this semi-tribal village, social stratification on the basis of landownership was less pronounced than in the average villages of Gujarat. The large majority of the people in the village belong to the Barya community. There are a few Tadwi and Naika households and only two Patel families. The social structure is very homogeneous making it easier to implement many activities for women in the village such as the smokeless chulha programme, a better system for drawing drinking water, kitchen-gardening, cattle breeding courses, etc. Heavy emphasis is put on winning the confidence and support of women and on interconnecting different aspects of change initiated by rural development projects which affect the working patterns of women.

### 11.3.2 Smokeless chulhas

In the context of this study it is important to point out that there was no perception of an increasing scarcity of wood fuel in this village. Even today landless families do not buy wood but gather it free from the forest. I was not able to investigate if this perception was in line with the real ecological situation. My impression based on observations in the forest area is that there is no indiscriminate cutting of trees by the villagers and that there is an awareness of the need to protect small trees. Nevertheless, the forest area is rather thin in some places, and the first signs of soil erosion are visible.

In 1982, a programme to build smokeless chulhas was introduced. The improved chulhas consist of mud enclosures over the fire with two pot-holes; the hole in front being bigger than the second one. There are no dampers or sliding doors in front of the firebox. To carry smoke away and to create a necessary draught, an asbestos chimney is placed through a hole in the roof covered with a cowl to prevent rain from entering the chimney. The project was implemented by students of Baroda college who have to select an additional voluntary subject in their final year which can either be 'sports' or 'social service to the people'. Within the framework of the latter option, National Students Service camps have been organised with government support to instruct students in construction of latrines, biogas plants and smokeless chulhas. These students are mostly boys, as parents do not approve of their daughters of marriageable age to take part in such activities where they have to stay somewhere in the countryside.

On the initiative of Dr. N., the leader and initiator of the whole project, boys started to build chulhas in the village. The first one was built in the house of the sarpanch (village head) who was also taught how to construct chulhas. This one became the model which was shown to village women. It seemed that all the village women were very keen to have this new cooking stove. For financial reasons it was only possible to meet half of the demand. Fifty chulhas were financed by a grant from the district panchayat. Dr. N. did not concern himself with the selection criteria. The streets around the house of the sarpanch have chulhas; the streets on the other side of the village do not.

The main objectives of this chulha programme are to reduce deforestation by saving fuelwood as a result of the expected higher efficiency of the new stove and to reduce the inconvenience of smoke.

### 11.3.3 Advantages and disadvantages of the 'new' chulhas

From interviews with women users in the villages it was clear that there were several advantages of the new chulhas: they are more comfortable, save time, are modern, cook food more quickly, and are easier to use. The new chulhas also produce less smoke thereby keeping the kitchen clean. When women claim that 'it cooks quicker', they mean that two dishes can be cooked simultaneously. When the rice is boiled, they can place the pan on the second pot-hole which gets enough heat from the first pot-hole on which 'roti' can be prepared.

It is significant that all these arguments indicate that women appreciate the saving in their cooking time as they now work more hours a day in their own fields than ever before. This is a consequence of the recent introduction of improved seeds and chemical fertilisers in the village. Before this agricultural development, the majority of households were self-sufficient in food production for six months of the year as dry land cultivation allows one kharif (monsoon) crop. For the remaining six months, women and men had to rely on wage labour outside the village. Often they had to walk 10 miles to earn Rs. 5 to Rs. 6 a day. Everywhere women informed me that because of the introduction of new agricultural inputs they now worked more hours a day in their own fields and the reward is an increase in production and income. Therefore, the need for wage labour has decreased considerably in the last two years. It is estimated that households are self-sufficient, in food production for nine months of the year. Women perceived these developments as a significant improvement in their lives.

My visit to the village coincided with an important stage of rural transformation from a combined self-sufficient/wage labour economy to one showing the first accumulation of capital surplus. This process occurred over a short period of five years. The surplus, which was the result of a 50 per cent production increase of some crops, was invested in chemical fertilisers and especially bullocks. Previously, peasants rented bullocks from a local landowner and had to pay in kind (up to 30 per cent of their crop).

The increase of labour time for men and women in cultivating their own fields has more consequences for women than for men because women are also responsible for household work. This helps to explain why at this stage of rural transformation, the desire to save time, particularly cooking time, is a very important motive for accepting the 'new' chulha. It is interesting to note that only one woman did not report any saving of fuel, and another one even said that more fuel was required.

It was not possible to affirm whether the high hopes that have been placed on stove programmes in general as a means of saving energy on a large scale and reducing deforestation have been realised. These hopes are mostly based on stoves tested in a laboratory. Field conditions are different and construction faults are common and can be expected of school boys building chulhas in a 'social service' campaign. The drawback mentioned by three women using an improved chulha is that no repair work could be done for the part of the chulha which was not under their control, i.e., the chimney. The main body of the chulha can be maintained in the traditional way: when it cracks women repair it using mud and cowdung. But when something is wrong with the chimney, or chimney cowls, they cannot do anything. In two cases, rain poured into the house because the hole around the chimney was too wide. So they broke the chimney and covered the hole with tiles. In another case, the cowl was damaged when the roof was being repaired. They had to stop using the stove because rain poured into the chimney. When there is no organised maintenance system under the control of women, problems arise.

### 11.3.4 Findings

On the basis of the foregoing we can say that the project is a success as summarised in the conclusion. This is also supported by the fact that women in 50 households who still used the traditional stoves were very eager to have a new stove. Women were not involved as directors, initiators, designers, constructors or maintenance workers in the project: men and school boys did the jobs. However, local women worked on a volunteer basis as demonstrators of the new chulhas in the project. They were well respected elderly women with grown-up children. Their role as demonstrators/propagators seemed crucial for the success of the project. The promotion of the stoves is an aspect of the energy project and a continuation of a rural development programme. The support and trust of women users has been won over a period of time and women are pleased that 'outsiders', men from the city with a high status, are interested in their cooking problems and methods. This attention has added value and importance to women's household work and has generated a lot of goodwill. The chulha programme is a new technological innovation which together with a number of others has been introduced in the village. To have a modern asset in the kitchen, a place where not much innovation usually takes place is something that pleases women very much. It seemed that the interlinking and integration with the total package played an important role in the acceptance of the stove programme.

As a preliminary conclusion we can say that the implementation of the stove project was quite successful because of the following conditions:

1. The villagers and especially women were mobilised and involved beforehand. In this way the distribution of chulhas was interlinked with a wave of changes which stimulated an open attitude to new options.
2. Women demonstrators of the chulhas are respected local women with a great deal of knowledge of local conditions and problems related to the subject.

3. The homogeneous social structure in the village contributed to the uniform distribution of the new stoves in one half of the village. There was no tendency for only the better-off to have chulhas in their kitchen.
4. Women themselves decided, to a large extent, to have a 'new' chulha as patriarchal authority was relatively absent.
5. A perceived advantage of the new chulha by women users was that it saved time. In a situation where women have to work longer in the fields a reduction in cooking time is welcome. The phase of rural transformation and the impact of that on women's labour outside the home appeared to be crucial for the acceptance and use of the 'new' cooking stoves.
6. Women like to have a new 'gadget', particularly one which focuses attention on the value and importance of their management of cooking fuel and cooking.
7. Apart from the chimney and cowl, the 'new' stove is not much different from the old one. It is not necessary to change the management of wood fuel or the mode of cooking. Adaptations when, for example, the pots do not fit the holes, can be easily made by women themselves. There is also no question of changing food habits or customs. The new stove is in fact an improved type of the old one.

Yet there are shortcomings including the fact that no woman was properly trained to construct and maintain a stove nor was any users' education imparted.

## 11.4 WOMEN, HEALTH AND SMOKE IN KHEDA DISTRICT

### 11.4.1 Nature, objectives and organisation

The first contact with the work of the NGO in Kheda district was fascinating. When I arrived at the central meeting hall, more than 150 women working as Village Health Workers were attending their annual meeting. Many village women came to the microphone to make a speech or to criticise the staff members: 'Yes you have everything on paper, but we have to do the work in the villages . . .' I was amazed by the self-confidence, firmness and determination with which these women came forward and spoke to such a large audience. The emancipated and democratic atmosphere of this meeting made me very eager and curious to know more of the way in which this voluntary organisation implements its programmes and, in particular, its chulha project.

In December 1982, the NGO's programme was established in 82 villages of the district, and by January 1985 it had already been extended to 152 villages; many of them in Borsad, Petlau and Anand talukas. There are plans to expand the programme to more than 300 villages in Kheda district. By 1985 the number of staff members had increased to 122. At the local level, hundreds of volunteers, particularly women, receiving some remuneration are involved in the work of the NGO. Maternal and infant care, and preventive and curative medicine constitute the programme's core action item, to be initiated first when a village decides to participate in the programme. Acceptance of the core item requires a participating village's commitment to the programme. The family annual membership fee of Rs. 10 has to be accepted. The village has to agree, by consensus, on the selection of the Village Health Worker (VHW) from among all adult females in the village. The Milk Producers' Co-operative has to agree to pay half the VHW's salary. The Milk Producers' Co-operative's Management Committee has to set up a Family Health Care Sub-Committee which will shape the programme at the village level.

The foundation funding the project is a secular and reformist organisation whose ideology has been influenced by Gandhian ideas and beliefs; institutional viability is even more important than financial viability. This requires the development of people, and that is the approach of the Foundation's integrated rural development strategy of working 'from the people up'. The organisation receives finances from local sources such as membership fees, milk producers' co-operatives, the National Dairy Development Board, as well as from UNICEF and the Overseas Development Administration of the UK.

### 11.4.2 Smokeless chulhas in the villages

The first village visited was a multi-caste village comprising Patels, Rajputs, Kumbars, Dheds, etc., totalling around 120 households. Patidars have more than 40 acres of land, but most of the Dheds are landless. Even in this well-to-do village, malnutrition among children in 40 to 50 landless families was observed. Cash crops are tobacco and peanuts. The infrastructure of the village looked impressive. Electrification was introduced in the village in 1972 and, according to the sarpanch, 90 per cent of the houses have an electric power connection which costs between Rs. 20 to Rs. 30 per month. There is a primary school up to grade 10 standard. The milk co-operative was established in 1970 and there are 165 shareholders (27 women).

In the house of a Rajput family there was a smokeless chulha which was working very well. The woman of the house mentioned that the advantage was that there was less smoke and the surroundings were cleaner. At the time she was interviewed, she was using tur stalks from her own field as fuel.

The Village Health Worker and the NGO fieldworker have propagated the chulha programme in the village. They call the village women to meetings to inform them of the advantages. If a woman is interested, they show her a demonstration model installed in one of the houses in the village. Besides this, they visit every house for mother and child care work, and during these visits they also give information about the chulhas. The chulhas are built by a mistri, a mason from a nearby village who was trained to make them at the NGO centre. He is supervised by the engineer of the NGO. This setting gives some guarantee that maintenance of the chulhas will not be a problem. The VHW, fieldworker, or mistri are well known in the village and they can be easily approached by the local women when something goes wrong. The Village Health Worker is a young woman selected by the NGO, from among five to six candidates with 10 standard education. Once every 15 days she goes for training to the centre of the NGO. To solve the transportation problem and to guarantee the safety of women, jeeps are used to bring the VHW to and from the villages. She earns Rs. 100 a month and works three hours a day as a VHW. The VHWs also visited the falia (ward) of very poor Harijans to inform them of the chulha programme. In one instance, a demonstration had been arranged in the house of the secretary of the milk co-operative, a Patidar family. It was the first time that Harijans had entered the house of a member of this community and there was no problem related to caste discrimination or segregation. In this way the chulha programme has contributed to reducing the traditional caste barriers between ex-untouchable and higher caste women.

In another village, a Muslim woman, trained to construct

smokeless chulhas, had constructed 75 of the 150 chulhas installed. Small houses for Harijans planned under the '20-point programme' had been constructed in 1984–85, which allowed the inclusion of considerable numbers of chulhas in the houses of the lower strata.

In one ward, the new chulhas were not built inside the house but outside on the verandas. The necessity to reduce smoke seems less relevant in these cases. Women have to provide the material for the chulhas themselves: cowdung, clay and straw. The costs of the construction are Rs. 10. Among the advantages women mentioned that the new chulha is less smokey and two dishes could be cooked at one time. A disadvantage was that it cracks but can be repaired in the traditional way. Firewood is scarce and one family had to buy 20 to 30 kg of firewood a week at Rs. 12 to Rs. 18. They were obviously interested in saving wood fuel but till now they had not noticed any saving. This family was rather well-to-do and they could afford to buy firewood but in other villages poor households were really suffering from a scarcity of firewood.

In another village, this scarcity was not due to deforestation but related to the power of the panchayat. In this village, the Patidars are the big landowners and power holders, the Rajputs are the middle peasants, and Muslims, Harijans and Adivasi are the poor peasants and agricultural labourers. It is a 90 per cent tobacco culture village but cattle also contribute to large fortunes. The village has a large plot of wasteland (gaucher-land) but it is under control of the panchayat. None of the poor villagers are allowed to gather firewood on this wasteland where the panchayat looks after the trees, cuts the wood and sells it to the villagers (20 kg for Rs. 10). The poor now have to buy this wood fuel for which, in the past, they had a communal right to collect free. In the Harijan ward women suffering from TB could only cook one meal a day because of lack of money to buy firewood. I was not able to find out when and in what way, the panchayat was able to obtain usufructuary rights to the wasteland. It is also in this village that severely undernourished children have been identified and taken to the children's hospital at the NGO centre in order to save their lives.

### 11.4.3 The success of the programme

In general, this chulha programme seemed a success which can be attributed to several conditions:

1. It is one item, part of a whole range of activities already known to the villagers, especially designed to improve the life of women and children of poor households.
2. The chulhas are properly constructed and maintained by local men and women and well-known women fieldworkers, who have already shown their sincerity in the mother and child care programme.
3. The organisation is headed by a spirited and dynamic woman who is a qualified doctor and initiator of the chulha programme.
4. Local workers have opportunities to influence the decision-making process by giving criticism or suggesting improvements.
5. The chulha programme and the mother and child care programme have generated employment opportunities for poor rural women. Moreover, in order to reduce the problems women face in being active outside the house, the NGO created the possibility for increased freedom of movement by providing jeep transport and access to knowledge and information (training and education). The project offered a unique chance to some women to develop their personality and capacities.

## 11.5 A MULTI-COMMUNITY VILLAGE, MEN AND CHULHAS

### 11.5.1 Village and Voluntary Organisation (VO)

The third project was in a village near Ahmedabad, situated in Sanand taluka, 25 km southwest of Ahmedabad. A sandy road leads to the village where huts and houses are scattered over a large area. There are electric transmission lines 1 km from the village but electrification has not yet taken place. The total population is 1,703 with approximately 330 families. The main crops are cotton, jowar, paddy and bajra. Animal husbandry is a supplementary means of subsistence. The village is an obvious multi-community-caste village with Hindu and Muslim populations. The Hindu population consists of many sub-caste groups living in separate wards: Durba, Thaharda, Bharwad, Rawbari, Dhed, Kumbar, etc. A survey of the organisation working in the village shows that there are 73 small and marginal farmers and 101 landless families.

The objective of the VO, which has installed only 10 smokeless chulhas in the village, is to work as a catalyst in the initiation schemes of the various government departments such as the district and taluka panchayats, and semi-government institutions. They try to obtain approval for schemes such as those to supply water and improve agricultural implements. They organise training courses for rural artisans such as handloom weavers, potters and blacksmiths. The trustee of the VO considers its role as one of assistance and guidance with local institutions carrying out the activities and not the VO which has adopted the village. The role is therefore more one of coordination, providing 'managerial inputs' at the village level. The objective is to improve the quality of rural life and provide employment opportunities to increase the income of the rural poor.

### 11.5.2 The 'new' chulha project

The smokeless chulha programme was implemented 'top-down', without any survey or investigation of needs or wishes of women in the village. 'One morning they came and brought the material', according to one woman. The participation of the village population in the project was absolutely nil. Two men, working as volunteers at the village level, were supposed to mobilise the people but they were unable to do so because of the many constraints in the social structure and cultural patterns. One of the volunteer workers is a Harijan. He and his colleague were not able to reach the other eight communities nor women of their communities because of caste discrimination and sex segregation. Moreover, it seemed that they have no notion of the problems women face in relation to the collecting of cooking fuel and cooking. The sex segregation notions of Rabari, Bharwad, Bhoi or Menon groups in particular made any communication with women of these groups and the two men promoters impossible. Women who had a 'new' chulha in their kitchen accepted it, as they had been told that it would save firewood, but they complained that they had to use more firewood than in the old stoves. No doubt there were construction faults in this case because the volunteers had not received proper training and were not at all equipped to maintain the chulhas nor to instruct women on how to use them properly.

### 11.5.3 Reasons of failure

This chulha programme is a failure for the following reasons:

— Technical failures.
— Inadequate training and maintenance structure and no users' education.
— No women are involved or included in the organisation responsible for installation or working at the village level.
— The nature of the VO not working directly at the grassroots level but only as a catalyst agent.
— A windmill project to draw water had also been a failure and affected women of the village in a very negative way. This experience discouraged women and made them reluctant to accept another women's innovation initiated by this VO.

However, it has to be acknowledged that even if many enthusiastic women village workers could be appointed to this programme, success would not be easy. The differences in classes, castes and communities are deeply rooted, making interaction and communication a significant problem. The strong patriarchal relations and the limited freedom of movement of especially Muslim women and girls were obvious. They were only permitted to receive religious education at the village Madrasa (Muslim school) and to learn Urdu. The majority of Muslim women veiled themselves while fetching water from the public well.

It is especially in these situations that women of a NGO or a voluntary organisation should work intensively, and directly at the grassroots level and increase inter- and intra-community participation.

## 11.6 THE KEY TO SUCCESS: SKILLED WOMEN CHULHA-MISTRIS

### 11.6.1 Village and NGO

The main centre of the NGO in Panch Mahal is not easily accessible being situated in a remote area in a new, simple building. The volunteers of the NGO were dedicated, serious and committed to helping poor villagers to develop themselves, while they bear hardships with the bare minimum to sustain them. The backbone of the NGO activities is drinking water or the Ground Water Development Programme. Other important activities are in the field of rural development such as providing crop loans and income generating projects such as cottage industries. People are trained in weaving, are helped to set up a weavers' co-operative, and looms are provided through the mediation of the NGO. The NGO also organises evening classes for children from 6 to 15 years of age who are not able to attend the formal day school because they have to herd cattle, do household chores or work in the fields.

In addition, the NGO installs improved chulhas (of the Nada-type) in the villages in Santrampur taluka. In two years 240 Nada chulhas have been installed.

### 11.6.2 How the chulha programme runs

A local village woman had attended a course conducted at the NGO centre and was trained as a 'chulha mistri'. She and the head mistri were able to construct a Nada-type chulha with a chimney, damper, baffle and sliding metal doors in half an hour. In one instance, the woman of the house provided the clay and straw mix required to make the 'body' of the chulha. The size of the cooking holes was made according to the size of the utensils used by the woman of the house. The only task done by the husband in the whole process was to saw a hole in the roof for the chimney. In other cases it seemed that women mistris also perform this 'unusual'—at least for Indian norms—job. The mistris received Rs. 10 per chulha and the cost for the woman user in this case was Rs. 15 and for Harijans Rs. 5. The rest of the cost is subsidised from funds the NGO receives from national and foreign sources.

In other villages I witnessed the follow-up and users' education work performed by mistris. The mistris earn Rs. 50 per month to follow up their work and to provide users' education on the use of dampers, sliding metal doors and when and how to clean the chimney. The emancipation of women, which is promoted by the diffusion of chulhas, is a very important aspect of the project. Not only the income generating facet is significant but self-awareness is raised and poor women begin to rely upon their own capacities and skills. Of the 10 women mistris trained and working, two are Harijans and two are of tribal origin.

In the whole programme it is emphasised that simple demonstrations of the construction method is not enough. The training programme should be for at least seven to 10 days in which the trainees learn to handle real-life problems. There was one woman chulha supervisor who selected village women for training and allocated the various tasks to them. She was also involved in the advanced preparation of chaapris (the basic 'body' made in standard moulds).

In the programme it was also stressed that the chulha must meet the user's needs. Some housewives in Panch Mahal, for example, have strong views about the direction in which the chulha should face. The north-south position is believed to be the best and not the east-west. Other considerations are, of course, cooking fuel, cooking purpose, cooking routine, etc.

It is acknowledged by the centre that the chulha is unsuitable for extremely poor families who can only afford a one item meal. Some underprivileged tribal groups living in the area have other priorities in their lives than chulhas. Also, as they move frequently the fixed Nada-type chulha is obviously not useful.

In the area, there were no problems of women rejecting chulhas because there was no longer any smoke for the preservation of grain, seeds, meat and fish as well as for the destruction of insects. On the contrary, women said they liked the chulhas because of the reduction of smoke. A Harijan woman said jokingly 'my "bungalow" remains much cleaner. I save time and energy because the wall plastering has been reduced'. The argument that the chulha cooks more quickly scored here as the second advantage.

It can be stated that this chulha programme is a real success. The involvement of women in different functions is very clear. As most villagers are of semi-tribal origin, women enjoy more freedom of movement making it easier to mobilise the local women working at the grassroots.

## 11.7 FAILURE OF THE FORESTRY DEPARTMENT CHULHA PROGRAMME

In the same area where the NGO works, the Forestry Department also installs 'new' stoves, but our investigation and other information indicates that these chulhas are not becoming popular. In one year, 25 chulhas were introduced in five villages.

On the basis of a short investigation the main reasons for the failure perceived were:

1. A top-down approach had been used. The range officer and his watchmen, all men with no experience of cooking or collecting fuel, install the chulhas. Women are assigned a passive role, being expected to accept the chulha as something good for them free of cost.
2. The construction is carried out by using a manual. It is assumed that reading instructions is sufficient for the construction of the chulha.
3. The assistants are badly trained and do not get any material incentive to do their best.
4. There is no institutional arrangement for the education of users, repairs, maintenance, etc.
5. Those implementing the programme are not interested in the development and emancipation of the people.

In fact, all conditions present in the Nada stove programme are absent here.

## 11.8 RESULTS AND REMARKS

My main proposition was that the 'new' chulha programmes would be more successful when local women participate in various functions in the organisation installing the stoves. On the basis of the previous description and analyses of five smokeless chulha projects, Table 11.1 provides an overview of the gender of those fulfilling seven functions in the five projects studied.

Table 11.1 indicates that in two projects (Ahmedabad and Panch Mahal II), no women are engaged in any function. In one

TABLE 11.1
**Gender of Employees in the Management of Cooking Stove Projects**

| *Management Functions* | *Project Locations (Districts)* | | | | |
|---|---|---|---|---|---|
| | *Baroda* | *Ahmedabad* | *Kheda* | *Panch Mahal I* | *Panch Mahal II* |
| Director | M | M | F | M | M |
| Initiator | M | M | M/F | F | M |
| Propagator | M | M | F | M/F | M |
| Designer | M | M | M | F | M |
| Constructor | M | M | M/F | F | M |
| Demonstrator | M/F | — | F | F | M |
| Maintenance Worker | — | — | M/F | F | — |
| Total Functions | 0.5M / 0.5F | 5M / 0F | 2.5M / 4.5F | 1.5M / 5.5F | 6M / 0f |

project, women are only demonstrators of new stoves together with men (Baroda). Men dominated all other tasks. In two other cases, women accounted for 4.5 and 5.5 functions and men for 2.5 and 1.5 out of a total of seven. The success of the five projects, assessed on the basis of interviews, group discussions, observations and perceptions of the women users was as follows:

TABLE 11.2
**Assessment of Success of Project**

| *Project Locations (Districts)* | *Assessment of Success* |
|---|---|
| Baroda | Successful |
| Ahmedabad | Complete Failure |
| Kheda | Successful |
| Panch Mahal I | Very Successful |
| Panch Mahal II | Complete Failure |

The relationship between the male/female ratios and the success of the project is thus as follows:

TABLE 11.3
**Predominant Male or Female Involvement and Success of Project**

| *Project Location (District)* | *Involvement Management Functions (M/F)* | *Assessment of Success* |
|---|---|---|
| Baroda | Almost Male Dominated | Successful |
| Ahmedabad | Male Dominated | Complete Failure |
| Kheda | Female Dominated | Successful |
| Panch Mahal I | Female Dominated | Very Successful |
| Panch Mahal II | Male Dominated | Complete Failure |

Conclusions from this table should be drawn very carefully because, as mentioned previously, other variables may play a role. Nevertheless, the outcome gives a strong indication that the role of women seems to be of much importance. In four out of five projects my main proposition seems to be verified: the two male dominated programmes are a failure and the two female dominated ones are a success. Moreover, a more differentiated analysis of the importance of each of the various functions performed by the local women can help to explain the success of the almost male dominated project. Women are only chulha demonstrators in this project. Nevertheless, this function is crucial in view of sex segregation and perceptions of women's domain. This does not mean that the other variables for success mentioned previously are insignificant but it seems worthwhile to assess the importance of these factors in view of the findings presented here.

The study also reveals that a differentiated analysis of the various functions performed by women shows that it is wrong to speak of 'women' in general in the context of the chulha organisations. The class/caste differences between women in different functions are obvious. Urban middle class women are directors, initiators and designers; poor local women are involved in less rewarding jobs which require less formal education and have less influence.

Finally, we have to consider the impact of the two background variables: the level of socio-economic development of the district and the nature of the chulha organisation. There is no clear correlation between the participation of women in different functions in the chulha organisations and the level of development. In both the highly developed and 'backward' districts, the involvement of women is high. Tribal women live in the less developed areas. The status and position of these women, however, is often better than that of women living in the highly developed Hindu villages. Traditionally, tribal women have more freedom of movement particularly when they are married with grown-up children. In the highly developed district of Kheda the participation of women is also high although they are mainly Hindu. In this case, the chulha organisation has taken many precautions to avoid problems and to protect women. Jeeps are used to pick up women engaged in the organisation when they have to travel to the NGO or any other place, while in the work environment segregation of the sexes is ensured as far as possible. Construction work, which requires contacts with men, was undertaken by an elderly divorced woman whose position was, no doubt, atypical.

The study showed that the participation of women is likely to be higher in the chulha organisations which work directly at the grassroots level.

The final conclusion and recommendation is that in improved chulha projects women should be involved at *all* levels in the chulha organisation, including the higher ones. They should be trained in construction techniques, provided with adequate users' education and made highly aware of the socio-economic, cultural and political constraints.

# 12

# THE SUCCESS OF MIXED MOTIVES: STOVE AND FORESTRY PROGRAMMES IN GUJARAT[1]

**Margaret M. Skutsch**

'Success' and 'failure' in rural development programmes are not merely relative terms; they are fundamentally personal judgements, depending for definition on criteria which are group specific. The sponsor, the implementing agency, the recipient and the intended target group may all have different ideas about what is successful and what is not. Outside observers may hold yet other views. Thus a programme for stove dissemination might be considered successful by an international sponsor if it reduces the amount of wood fuel used by the household in preparing food; to the agency in the field, it might be considered successful only if it lightens the burden of work that women normally perform; on the other hand, it might be that the user would only consider it successful if the food can be cooked more quickly than before. An outside observer, such as a researcher on the IDPAD programme, might consider it successful only if it were able to reach the poorest sections of society and bring benefits specifically and differentially to this group.

The difficulty of defining success and failure is compounded by the fact that the officially expressed objectives of a programme, as formulated by the sponsor or by the implementing agency, are often only formal, or vague expectations of what the programme is supposed to achieve. In practice, the programme is often evaluated not on the achievement of these objectives at all but on some quite different and simple criteria. As will be noted in the case studies described in this paper, it is common for a sponsor whose official aim is to reduce the amount of firewood used in cooking, to use the number of stoves constructed during the course of the programme as the sole indicator of success. An agency interested in increasing the supply of firewood in rural areas may measure success in terms of the number of trees planted as a result of its work. The significance of the difference between official objectives and working indicators of success can be easily demonstrated in the cases studied in Gujarat; for it is apparent that the stoves introduced in many cases are no more energy efficient than the traditional fireplaces they replace and the trees which have been so widely planted are not destined for the domestic hearth at all but for the construction industry in the towns and cities. Yet these same programmes are considered highly successful by their organisers.

In this chapter, an attempt is made to define the formal objectives and actual criteria of the various parties involved in stove and forestry projects in Gujarat. It draws heavily on two working papers prepared during the research under the IDPAD programme (Skutsch, [1986, 1987]). In these papers, the emphasis was to explain the technological options and the dissemination strategy chosen by the programme managers in terms of the real concerns and the evaluation criteria held by them. In this study, attention is focused on the comparison of objectives of the various parties involved in each programme. The paper attempts to show that a programme is 'successful' not because the various parties involved shared the same fundamental objectives but because the programme is in some way able to satisfy their different objectives at the same time. The paper also highlights the difference between official objectives and management practice in programme assessment, and explains these differences in a rational framework.

The value of this kind of analysis is that it gives some general insight into the need for ex-post evaluations and highlights the need to keep in mind not merely the original objectives and targets of development, but also the values, goals, and constraints of all the key parties involved. It shows, for example, that monitoring of development programmes should be flexible and responsive to these factors. Moreover, those responsible for project design should be aware of the likely variations of goals within the hierarchy if they are intent on achieving success according to some pre-defined criteria.

## 12.1 STOVE PROGRAMMES IN GUJARAT

There are at least 20 different ongoing programmes which promote improved stoves in Gujarat. A variety of different sponsors, intermediate government agencies and implementing agencies, most of which are non-governmental organisations (NGOs), are engaged in this activity. The structure of a number of these programmes is presented in Table 12.1.

The Safai Vidyalaya Institute (SVI), which has close connections with the Harijan Ashram in Ahmedabad, is a key actor. This group has been involved in distributing stoves, together with sanitary hardware, for 20 years. The SVI works through the medium of camps of student volunteers who visit villages for a two-week period and construct a two-holed stove with a chimney

TABLE 12.1
**Structure of Eleven Stove Programmes Operating in Gujarat**

| *Sponsor Agency* | *Intermediate Agency* | *Implementing Agency* |
|---|---|---|
| Social Welfare Department | District Panchayat House Improvement Scheme | Safai Vidyalaya Institute (SVI) |
| Health Department | District Health Department Rural Sanitation Scheme | Safai Vidyalaya Institute |
| Khadi and Village Industries Commission (KVIC) | KVIC | KVIC |
| World Bank | Forestry Department | Forest Guards |
| Department of Non-Conventional Energy Sources (DNES) | Forestry Department | Self-Employed Women's Association (SEWA) |
| DNES | Gujarat Energy Development Authority (CEDA) | Tribhuvandas Foundation |
| DNES | CEDA | Prayas |
| DNES | CEDA | Sewa Jughadia |
| Oxfam | | Social Work and Research Centre (SWRC) |
| Society for Promotion of Wasteland Development | | Mahiti |
| UNICEF | Social Inputs in Area Development | Safai Vidyalaya Institute (SVI) |

following a standard pattern which they have been taught at the camp. The Rural Sanitation Programme, under the State Health Department, finances much of this activity. As a result of the good reputation built up in this field, the SVI has been asked to act as the implementing agency in a variety of other programmes with different financial sources, as Table 12.1 clearly shows. The stove built, however, is not always the same.

In some of the house improvement projects most stoves are built by trained technicians of the SVI, from good quality concrete as the funding is more generous from the Welfare Department. The Safai Vidyalaya was also responsible for training stove builders in a variety of other programmes, for example, that of the Self-Employed Women's Association (SEWA) and that of Sewa Jughadia, a small NGO working in Baruch district. Both groups work with the standard stove model. The Forestry Department and the Khadi and Village Industries also use the design and manuals of the Safai Vidyalaya in their programmes. These programmes rely either on volunteer camps for implementation, or on a pyramid training system, i.e., on stove builders who are temporarily employed for the purpose and once trained go on to train a second and even third generation of stove builders. In this manner the link with the original organisation and the stove builders in the field becomes rather distant, tenuous and uncontrolled. In all these programmes, a standard stove model is built, usually of prefabricated parts, i.e., a fixed number of bricks and a pre-cast top. The stoves built under these conditions are frequently found to be deficient by the users, and there is a high rate of breakage or discontinuance of use.

In contrast there are a number of organisations such as the Social Work and Research Centre (SWRC) in Panch Mahals, the Tribhuvandas Foundation in Anand, Prayas in Baruch district, and Mahiti in Dholera where stove builders are permanent employees of the implementing agency. In these cases the stoves are not generally built to a standard design but follow general guidelines and are adapted to individual household requirements. Generally, these stoves have a high survival rate and their users claim that they are very good. These programmes, however, are very small in comparison with the programmes of the first type, because supervision by highly qualified staff is involved.

## 12.2 SOCIAL FORESTRY PROGRAMMES IN GUJARAT

In contrast, the structure of social forestry programmes is relatively simple. Here the chief actor is the State Forestry Department which began to plant trees on land outside the normal forest boundaries in the early seventies. The Department started with roadside strip plantations; later introducing farm forestry, whereby farmers themselves plant trees on their own land; and panchayat forestry, in which a communal village woodlot is planted on common land. Other schemes included the 'social security through tree plantation' scheme, under which landless and tribal people in particular were allotted degraded forest land and paid a regular monthly wage to plant and maintain the land in the initial years, and the Malki Lands scheme in the Dangs under which tribals were given loans to compensate for the income lost when they permitted the Forestry Department to plant trees on their own farm land. Since 1980, the Community Forestry Programme, as it was called then, has received major funding from the World Bank and has been rechristened the Gujarat Social Forestry Programme.

In addition to the direct work of the Forestry Department, a number of NGOs have taken up social forestry. Many of these NGOs base their efforts partly on facilities provided by the state programme, i.e., particularly the supply of free seedlings from the state nurseries. The efforts of Anand Niketan Ashram are well known in this respect, but others like Mahiti are operating in very difficult areas such as the saline wastelands of Dholera where they are making slower and less obvious progress. The SWRC launched a social forestry programme in 1984 with its own nursery but due to various problems has not done much outside the project headquarters, while the Tribhuvandas Foundation has plans to launch a similar programme, but has not yet done so. These organisations work in relatively limited areas. There is a state-wide NGO, the Sadvichar Parivar, primarily a charitable and relief agency, which initially planted woodlots around religious shrines, and has continued through the medium of schools and educational institutions to encourage large-scale tree planting. In all these programmes there is a tendency to rely on a number of well-known tree species, particularly eucalyptus and other exotics provided through the state programme. Mahiti has, however, branched out with experiments with lesser known species and some NGOs also try to introduce fruit trees. The NGOs are involved primarily in farm forestry, although as noted some involve schools as well.

## 12.3 OBJECTIVES OF THE STOVE PROGRAMMES IN GUJARAT

Interviews with personnel connected with the stove programmes have been combined with statements appearing in reports to provide an overview of the major goals of these programmes. It is clear that these objectives not only vary from programme to programme, but even within any one programme, the views of the different management levels, sponsor, intermediate agency, and implementing agency, differ.

### 12.3.1 The sponsoring agencies

Sponsors fall into two groups as far as major objectives are concerned. The first group, which includes the Social Welfare Department, the Health Department and UNICEF, considers improved health to be the chief purpose of their stove programmes. Smokey conditions caused by burning biomass inside houses is known to give rise to respiratory diseases and unhealthy conditions. Indeed, all the early programmes were initiated with this as the chief aim. More recently, however, fuel economy has appeared as another objective. Agencies such as the World Bank, the Department of Non-Conventional Energy Sources (DNES) and, to some extent, Oxfam have this as the major aim in their programmes. They differ slightly in the reasons for desiring fuel economy; for some, the emphasis is on protecting the forests from unnecessary cutting of live trees, for others such as Oxfam, the emphasis is on reducing the work load of women in collecting firewood. Generally, both aspects are mentioned and are seen as complementary.

All the sponsoring agencies, however, actually assess their success primarily in terms of the number of stoves which have been built. Thus, this is taken as the indicator of achievement of all the objectives just mentioned. In no case was it observed that the sponsoring agency had taken pains to test whether the stoves once built were (*a*) less smokey, or (*b*) more efficient in practice than traditional fireplaces, far less whether they were actually responsible for a decrease in the incidence of disease, forest conservation, or reduced working hours for women. As far as fuel efficiency is concerned, reference was made in some cases to laboratory tests carried out prior to the selection of the stove model. No pre-programme tests of any sort had been undertaken to assess the impact of chimneys on the selected stoves in reducing smoke pollution.

### 12.3.2 The intermediate agencies

The intermediate agencies are mainly state or district level government departments whose major task is to identify and to oversee the work of the implementing agencies. Most of these intermediate agencies, by and large, share the objectives of the sponsors, although they tend to express a broader range of objectives. For example, the district panchayat House Improvement Scheme (HIS) is specifically directed at providing housing for landless and tribal people in villages and thus aims to redistribute resources. Its stove programme is deliberately targeted towards assisting the lowest income families to obtain an improved stove. The Gujarat Energy Development Authority (GEDA) expresses quite a number of objectives in its promotion of stoves, including both smoke reduction and fuel economy. Another interesting example is the Forestry Department. Although one objective is fuel economy, it is clear that its working objective is quota achievement as, in this programme, there is strong pressure for accountability introduced via the annual reporting and budgeting system. In practice, the motivating factor appears to be to build the number of stoves specified in the year plan, some officials being quite clear on this. Both GEDA and the Forestry Department have been nominated as 'nodal agencies' for Gujarat by the sponsor, the DNES, although usually only one nodal agency is allowed per state. The Forestry Department was nominated because of its experience of working with the World Bank. It is a clear case demonstrating rivalry between intermediate agencies, which probably also exists in more covert forms between the other government bodies, for all of whom participation in a programme of national interest is important for long-term recognition and strength.

In all cases, the indicators of achievement at this level are the number of stoves built. The intermediate agencies had not undertaken any further studies to assess whether the stoves built had achieved the formal objectives.

### 12.3.3 The implementing agencies

The implementing agencies are all, except one, NGOs. Their objectives in working on the stove programmes are in general much broader than those of the bureaucratic agencies above them. Most of them claim a variety of objectives. The reason for this must be analysed carefully. These agencies are in a precarious financial and political position, at least compared with government agencies; they depend for much of their finance on political goodwill. Thus, each tends to outwardly express a large number of objectives to spread the risk, as it were. In particular it can be noted that these small NGOs are sensitive to the changing political climate and will adapt their own objective to those which enable them to command power and money at the moment. Thus, the whole issue of rural energy problems and the need for fuel economy is formally espoused by the NGOs but their officers, when pressed, often privately indicate that these objectives are of secondary importance in their view.

Objectives which do seem to motivate the implementing agencies are often health related ones; cleanliness and smokelessness. But there are other unwritten objectives too which come up during the course of a conversation. These include employment creation for stove builders, particularly for women. In the case of the SWRC, the basic aim of the programme was indeed to promote emancipation for women by creating a new women's profession. SEWA also perceives employment of women as a major objective, together with reducing rural women's burden of collecting fuel. Another objective of this kind is to provide the agency with entry into the villages. Some of the implementing agencies perceive the stove programme as a good starting point for developing relations in the villages where they wished to work, since stoves are relatively cheap and simple and are potentially of value to all households, in theory at least. Tribhuvandas Foundation, for example, felt it could use stoves to build up popular support for other interventions which the organisation considers intrinsically more important. Prayas and Sewa Jughadia had similar approaches. Some agencies even mentioned that money was easily available for stoves and the unit subsidy was so high, that funds could be saved and actually diverted into other programmes which are not at the moment popular among the funding agencies. This became a strong incentive for taking up stove building.

A special case among the implementing agencies are the Forest Guards. The main objective of those interviewed appeared to be achievement of their target and collection of their personal stipend, i.e., Rs. 5 per stove.

Despite the broad and interesting range of objectives, the indicator most often presented in the implementing agencies' own evaluation of their stove programmes was the simple one of the number of stoves built. A few, however, were more self-critical and they tried to ascertain the number of stoves that were still in operation after a given period of time. Prayas, for example, took the trouble to undertake this exercise, and the Safai Vidyalaya had some figures on this aspect in an internal report even if they were not entirely reliable and were not quoted publicly. Some agencies were interested in the number of jobs created, especially for women, and quoted this figure along with the number of stoves built. None of the agencies had figures on how many stoves had been built for poor people as opposed to rich people, except in the programme under the HIS where all the stoves were for poor people by definition. None of the implementing agencies had made any scientific studies concerning the extent to which smoke is reduced in chimney stoves, nor an analysis of whether health had improved since their introduction. No such reports appear to have been consulted in the original choice of technology either as none were known to exist. Only one agency (Tribhuvandas) had made field assessments of the efficiency of stoves in practice.

### 12.3.5 The users

The literature is full of hypotheses about what rural women expect from new stoves, including smokelessness, speed of cooking, fuel economy, relief from mosquitoes, preservation of the thatched roofs, etc. What is clear is that the class differences and thus the differences in requirements of rural women are tremendous. Cooking habits range from 'vegetables-in-rotis', cooked fast over a very hot fire, as in the case of poor women, to the rich women's assortment of dishes including rice and slowly cooked dals. The wealthy families, with a much more varied diet, need the capacity to control heat during cooking. Moreover, access to different sorts of fuel depends on class and caste. Rich families obtain nearly all their fuel in the form of wood from trees on their own farms, and thus they do not face any shortage. Poor women often use low grade crop residues and have much more trouble obtaining firewood when they do use it.

On the basis of the interviews with women of different classes in a number of villages, however, it was clear that cleanliness and smokelessness were qualities looked for in a stove by a broad range of women. Many women were using improved stoves that actually consumed more fuel than their traditional chula because of these qualities. The fuel economy criterion was more important, as might be expected, among poorer people. The appearance, the 'modern look', was certainly sometimes important. However, the key factor, as far as could be determined in these areas, was robustness of the stove, and whether it could easily be repaired by women themselves in case of cracks or breakage.

The indicators used by women in assessing their stoves were in terms of how long it lasted before cracks appeared; whether it could be repaired; the number of times they had to clean the kitchen walls; the number of bundles of firewood used per week; and an assessment of the quality of the air in the kitchen.

## 12.4 THE OBJECTIVES OF THE SOCIAL FORESTRY PROGRAMMES IN GUJARAT

### 12.4.1 The sponsoring agencies and intermediate agencies

In the social forestry field it is not possible to separate the sponsoring activities of agencies from their intermediate supervisory function and hence they are considered as one. The first agency to sponsor social forestry activities in Gujarat was the Forestry Department. The objectives were at first quite clear; a directive had been issued from the central headquarters that each state should have 33 per cent of its land area under trees, but in Gujarat only 12 per cent of the land is officially classified as forest and even on this land large areas have been deforested. The only solution was to grow trees on non-forest land which was justified in terms of supplying wood fuel to relieve pressure on the forests, and providing the villagers with timber. When the World Bank became a major sponsor, the objectives changed slightly to include employment for landless people. In addition to these formal objectives, however, there is undoubtedly an unwritten objective which relates to the prestige and reputation of the sponsors. Gujarat is a pioneer case of social forestry and both the major sponsors have obviously much to gain from showing the world their success.[2]

There are, in addition, a number of minor projects which are sponsored internally, such as the programme undertaken by Anand Niketan, in which it is difficult to distinguish the motives of the sponsors from those of the implementor. An interesting case is the Mahiti programme which is financed by the National Society for the Promotion of Wasteland Development. This is an environmental agency concerned mainly with erosion problems and thus interested in producing fuel through social forestry in order to save the existing forests.

The indicators used by those in all the agencies interviewed were related to the number of trees planted, or estimates of area planted per unit time, based on the number of seedlings recorded as taken from the nurseries. To a lesser extent, the number of farmers involved was also tabulated. The World Bank sponsored programme had made an attempt to collect data on the distribution of trees among rich and poor farmers but these records were badly maintained, as were figures on employment for landless and tribal people provided by the programme. A monitoring exercise was also carried out by the World Bank programme, but again this concentrated on indices of immediate output. No records were maintained by the few NGOs that were operating in this field. Comments made by officers in these NGOs, however, indicate that the number of seedlings taken was the major parameter considered.

It appeared that no studies had been undertaken concerning whether the wood was being used in urban or rural areas, or for what purpose, such as firewood, construction or pulp. The only study that had been carried out was a 'marketing' study which tried to make forecasts about whether a market for the wood would develop. There were no apparent plans to check any indirect results of the programme or whether the marketing study had been correct in its assessment.

### 12.4.2 The implementing agencies

The agencies under the state forestry scheme are the local forest range offices which cover sub-district areas. Although officers stationed here mentioned a variety of reasons for tree planting, it was apparent that they strongly supported the profit motive of farmers. At these offices, rough estimates of the expected market price for the wood were quoted, information which was not available in the head offices. Indeed, the range officers themselves gave the impression of being very well aware of the goals both of their bureaucratic superiors and of the farmers with whom they were working.

In general, small NGOs seem to be motivated either by strong environmental goals, such as Anand Niketan and Sadvichar Parivar, or by political reasons, such as Mahiti. In the case of the former, the aim is greening of the landscape for the harmonious balance of man and nature, and has strong spiritual underpinnings; in the case of the latter, social forestry was seen as a vehicle for building up the strength of a particular oppressed section of society against the local controlling forces.

In all cases, however, the indicators used are the same; the number of seedlings planted, although survival rates are also taken into account, as well as the number of farmers involved. No attempts were made at this level to determine systematically whether the trees grown were being used for firewood and by whom, and whether the programmes were successful in decreasing the use of forest wood; nor were any plans to make such evaluations found during the course of the research. On the political and environmental quality issues, the implementing agencies had only just begun their work and there was no way of really assessing their success. Most believed that success was, however, already on the way. Like many NGOs working on shoe-string budgets, the staff continued to work driven by a sense of inner commitment and energy rather than on the basis of scientific or systematic management analysis.

### 12.4.3 The users

In this case the 'users' of the technology must be defined as the 'adopters', i.e., the farmers and village communities who decided to grow trees under the scheme. Although most of these adopters are middle and large farmers with more than 5 acres of land, some marginal farmers with as little as half an acre were also interviewed. These farmers were growing eucalyptus trees around their farm boundaries. If anything, the profit objective was more important for the small than for the bigger farmers. Important additional aims include reducing their labour problem during the peak season, when labour has to be hired to harvest cereal crops, for example. The cultivation of trees, especially after the first year, requires much less labour than the traditional wheat, rice, millet and cotton.

A number of villages had also introduced self-help woodlots under the social forestry programme of the Forestry Department. While there is no doubt that these trees were intended for sale, to add to the panchayat funds, it was also mentioned that the purpose of the woodlot was to encourage individual farmers to take up tree planting themselves. In one village, the grass that grew well beneath the trees and was protected from grazing was made available free of cost to poor villagers to use as fodder, with permission of the sarpanch. This was stated as a secondary advantage of the scheme.

Few farmers had harvested trees at the time of the research, so it is difficult to say how the farmers and villages measured their success. Generally, all expressed great optimism about the large income they expected from the sale of their trees. Not a single one mentioned their use as firewood, except in a few cases when it was said that trimmings might be used by the family for this purpose; nor did the farmers' wives who were interviewed think that the trees were intended for this purpose.

## 12.5 SUMMARY OF MIXED MOTIVES IN STOVE AND SOCIAL FORESTRY PROGRAMMES

From the foregoing analysis it is clear that the reasons why users adopt new technologies differ from the reasons why sponsors originally promote them. Moreover, the implementing agencies have a different set of motivations which are quite unrelated to these other two sets of objectives. The intermediate agencies generally perceive things more in line with the sponsors. In other words, the programme can be simultaneously considered successful at different levels for quite different reasons.

However, several programmes were considered unsuccessful by the users of the technology. They were stove programmes in which a standard stove of prefabricated parts was distributed. The main reasons for their failure were a high breakage rate of the stoves and the fact that some of them used more fuel than the traditional fireplaces. Most women, however, did agree that the stoves were cleaner and less smokey. These programmes included those of

Prayas, most of those under the Rural Sanitation Programme (except those that were built in cement by professional masons), and those of the Forestry Department. There was no evidence of failure from the users' point of view in any of the social forestry programmes. However, many small farmers who would have liked to join the scheme were unable to do so because they did not own any land.

The implementing agencies had a different evaluation of these programmes, owing to the fact that the indicators they used did not reflect the concerns of the users. The only stove implementing agency which admitted failure was Prayas. This group analysed the reasons for the failures. It came up with another set of stoves of a different design in order to overcome the problems with the initial stoves. The other agencies, including the Safai Vidyalaya Institute and the Forestry Department, did not admit to any failure in their stove programmes and pointed proudly to the large numbers of stoves built. It is of interest to note that in this respect some NGOs work in the same manner as some of the government agencies. The difference is more in terms of scale than of institutional structure, procedures or attitudes.

The intermediate agencies and sponsors perceived every programme as successful. Some, such as GEDA, are working with a number of implementing agencies at the same time. The relative failure of one agency, Sewa Jughadia, to produce large numbers of stoves was taken together with the relative success of another, Tribhuvandas, to prove that the general strategy of working with NGOs was successful. In the social forestry programmes, the number of trees and hectares planted has been so astounding compared with the targets set, that there has been no cause for the sponsors and intermediate agencies even to consider that there might have been any failure involved.

Nevertheless, most of the stoves that have been built in the large-scale, prefabricated stove programmes have broken, and of the ones that are not broken, many save little or no fuel. All the wood produced under the social forestry programmes is being transported by dealers to the cities, where it is used for construction purposes or possibly sold to the townspeople as fuel. No trace is left to relieve the wood situation in the villages, and particularly not the wood fuel problem of the poor people who cannot afford to pay for commercial firewood.

## 12.6 CONCLUSION

Evidence from the fieldwork clearly shows that it is not necessary for all levels in a programme to share the same objectives for the programme to be successfully implemented. A programme must be flexible enough to take into account the particular needs and constraints of the various parties involved, if anything is to happen at all. The findings indicate that programme evaluation from an outside point of view must be sensitive to these varying requirements within the overall organisation, recognising the key constraints and expectations of the different levels.

The findings also show that some programmes that are unsuccessful from the viewpoint of the users can still survive and continue to obtain support and funding. It was seen that this was at least partly the result of using improper indicators for evaluation at all levels in the programme hierarchy. How can this seemingly irrational management practice on the part of major agencies be explained?

Many outside observers would immediately assume that it was the result of personal profit being made out of the funding system, from the village level upwards through the hierarchy. This aspect was not investigated in detail in the research, although incidental cases of corruption were evident. Certainly this may be a contributing factor within the organisations but it is not the central one.

Other observers would explain the situation to be 'yet another' example of systematic exploitation of the poorest groups in society by those with more power with the collusion of the major agencies. The guise of 'development projects' is taken on while the status quo is actually being reinforced. There is, of course, some universal validity in this interpretation, but it is not an adequate explanation.

A third avenue of explanation which might be followed is that of whether the management practices observed are, in the local context, more efficient than the 'objective' methods against which they are implicitly contrasted in this paper. These 'objective' methods of evaluation are very much anchored in a Western perception of what development programmes are for and how they ought to be run. Organisations in India have internal structures and are based on values which reflect Indian culture and social relations. The impact of these on the management methods used is not always openly discussed, because the Western model is often implicitly considered superior. One reason for this is that so much of the Indian code of practice rests on the cultural system of claims and obligations which at least in the idealised Western model are absent. In practice, the phenomena of mutual backscratching and old boys' clubs are more or less universal in development projects all over the world just as they are in business. Yet, in a general sense it would be worthwhile to investigate in more detail how management practices reflect the particular idioms of the Indian culture. It is likely that the observations made in the field are more fundamental and reflect behaviour patterns of development organisations in general and practices which might be recognised in many countries.

Therefore, another set of explanations is proposed, which rest on the understanding that the real purpose of any programme is its own existence, and not its real output, direct or indirect, in terms of objectives or indicators. The existence of the programme is important to all levels in the programme structure, in terms of funding, personal careers, and the position of the various agencies, particularly the intermediate agencies, with respect to competitors. The true purpose of a programme is to survive, and to strengthen the position of the various parties involved in the programme in the political arena.

A simple measure of success such as trees planted, or stoves built, is thus quite adequate for the real needs of the parties involved. Detailed strategic monitoring of the long-term results of the programmes would involve an exercise which would be relatively difficult from a technical point of view and take considerable time and effort away from the implementation itself. Furthermore, it is difficult to make 'fair' comparisons if complicated indicators are used. The use of a simple criterion restricts possible internal criticism or accusations of favouritism. A simple index is universally understood, while complex ones are subject to alternative interpretations and disagreement. Technical and management-psychological factors favour a straight output index and discourage the unnecessary complications of a more far-reaching and perceptive evaluation, from which none of the concerned parties in the organisation of the programme stands to gain.

## NOTES

1. This paper is based on fieldwork carried out under the IDPAD programme in February-March 1984 and January-February 1985. Many thanks are due for assistance from Professors T.K. Moulik and T. Gupta of the Indian Institute of Management; also to Dr Loes Schenk and Ms Sheela Thanki for advice, insights and companionship in the field. Very useful comments were made on the first draft of the paper by Professor Enno Hommes, Dr Mike McCall and Dr Jas Gill.
2. Here social forestry is considered to be forestry by the people, thus including self-help village schemes and farm forestry by individuals. Other elements of the Gujarat Social Forestry Programme involve a large component of government finance through, e.g., paid labour, and are therefore excluded from consideration.

# 13

# CONCLUSION: MANAGEMENT ISSUES

**Enno W. Hommes**

In the introductory chapter on management and the socio-cultural context it was pointed out that there are at least two schools of thought on how to evaluate rural technology projects. The anthropological school concentrates on the socio-cultural context and tries to explain success and failure in terms of concepts derived from the socio-cultural mechanisms in society that affect the projects. The environment of the project not only defines the scope of the project but also moulds its management. Management styles are defined by cultural codes. The second school, derived from business administration studies, concentrates on the different aspects of management itself, culminating in the concept of strategic management. According to this school, success and failure of projects can largely be explained in terms of good and poor management. Going through all the case studies presented, the main point of discussion remains undecided, but the value of this volume lies in the fact that it provides a unique collection of analyses which further our understanding of both points of view. As mentioned in the introduction, the value of the case studies is reflected more in the experience of the researchers than in the common methodology. This again makes it difficult to draw conclusions. Nevertheless, I feel the material justifies the conclusion that both the anthropological and the managerial approaches contribute to an understanding of projects in their totality. At the same time, the conclusion seems to be justified that there are few examples of case studies where the researchers are able to combine the two approaches effectively. The conclusion that good management should take the socio-economic and cultural environment into account is easier to draw than to apply. The authors did not come across projects which could be considered as having ideal approaches. What, in fact, would constitute an ideal approach? In any case, it should include the skilful application of all modern management tools which support the different steps in the dynamic project cycle. This means conducting surveys, need assessment studies, technology assessments, feasibility studies, monitoring programmes and evaluation studies.

In addition to this managerial structure, the management should take the socio-economic and cultural environment into account and translate this environment into project strategies and tactics. If the case studies are considered as contributions to an exploratory study then a new hypothesis can be formulated: the more aspects of the ideal management model covered in a project, the more successful that project will be. Further research is needed to indicate which aspects of the model are the most strategic.

After having analysed the different case studies, I feel that more approaches are possible. I would like to make some observations on at least three of these which can be termed the technological, the institutional and the 'span of control' approaches, respectively.
approaches,respectively.

### Technological approach

This is an approach in which a number of parameters are more or less determined by the relevant technology.

The first of these is the technology push. In quite a number of projects, a particular technology became available which led to the formulation of a project to exploit the technology. The final users were not consulted at all. Windmills are a good example of this sort of project. There can be many reasons for the technology push which must be traced back to interests groups. For example, politicians who want to spread the technology faster than is justified by the state-of-the-art, which is what happened in the case of windmills in India. There are also commercial or scientific interests that push a technology. In some of the case studies, the choice of technology was made at the international level and implemented through the national to the local level. Health projects are an example.

A second issue is the immaturity of the technologies concerned. One would expect a carefully planned sequence of fundamental research, the development of prototypes, testing in pilot projects, and, finally, the implementation on a large scale of the technology as tested. However, this often does not happen in practice. The price to be paid is technical failure. Windmills and stoves are two examples, but it would be easy to find many more examples in other fields.

A third issue, related to immaturity, is the fact that the users are first time buyers. The market is still to be developed and the users' experience is very short while a distribution system has usually not been developed. Proper attitudes toward operation and maintenance are also to be developed.

A fourth issue is the technological constraints of the system within which the new technology must function. The cold chain problem is an example. Without the existence of a cold chain many technically based medical solutions must be excluded. The level of skills required for the maintenance and repair of the support system also falls into this category.

A fifth issue is the market segment for which the technology is appropriate. In all case studies these market segments differ. A proper identification of the market segment is necessary. Biogas installations are neither for landless labourers with no cattle nor for farmers with too few cattle.

Another issue, related to the market segment, is that some of the technologies in our case studies are site specific and therefore must be almost tailormade. Windmills and stoves are examples. This site specificity makes these technologies unfit for regular marketing channels. They can only be installed with much additional effort.

These six issues together should form part of a technology assessment in which a kind of SWOT (strength and weakness) analysis can be applied. Technology assessment presupposes a management structure in which technology assessment can be applied. Unfortunately, there is often no such management structure. For a technology assessor the situation is different. He is in a position to apply a SWOT analysis and this will contribute to our understanding of success and failures.

These six factors explain why these technologies are not commercialised or only partly commercialised. At least in the initial stage, the government or NGOs have a role to play. Some technologies such as stoves and biogas installations can be commercialised after a period of time. For others, for example, in the medical field and infrastructural services such as drinking water and electricity, no road to commercialisation is yet evident. It is interesting to note that commercialisation is generally not part of the project strategy in the case studies. Where commercialisation has taken place, for instance, in energy technologies for decentralised energy production, this production is a sideline for the companies concerned. Jyoti in Baroda is an exception.

**Institutional approach**

The institutional approach is a method of analysis in which the institutional setting, e.g., bureaucratic institutions, political institutions and NGOs, takes a central place. With their own structures, procedures and cultures, they form the institutional environment within which a project must operate. The project can influence them to only a limited extent whereas they influence and sometimes even determine the nature of the project. However, they are at the same time both an asset and a constraint.

A major institutional constraint is the result of administrative and political decentralisation and the decision-making structures linked to this decentralisation. A rural water supply project in Gujarat has managerial constraints because some of the strategic management decisions are taken in Delhi. Choice of technology, financial structures, pay scales for personnel and many other details are not determined by the project management.

A very serious implication is that many management tools such as base-line surveys, need and technology assessment studies, feasibility studies and input-output analyses cannot be applied. The use of these management tools is common in industrial projects, especially in the private sector, but they are difficult to apply because the administrative and political level at which decisions are taken is not the level at which the project is implemented and where the project management is located. The project management cannot be blamed for lack of flexibility. In general, the distribution of power in the cases that were studied makes management extremely difficult. Sometimes it is even difficult to identify where power lies. The pluriform power structure in and around a project is also reflected in the pluriform objectives of the interest groups concerned. Margaret Skutsch's article on stoves provides a clear example.

Rural technology projects have to make allowances for government institutions, political institutions, manufacturers with their own organisational structures, banks and non-governmental organisations. The leading agency, which differs from project to project, has to deal with the problem of controlling the other parties concerned which can only be done to a limited extent. Coordination is necessary but not all the parties concerned cooperate. Sometimes this problem has been recognised by the government and new institutions, such as the GEDA in the field of energy in Gujarat, have been set up to solve it.

It could be asked whether government institutions are suitable organisations for implementing projects as they were originally designed to perform duties such as collecting taxes and maintaining law and order. Only with the advent of the development function was the government required to deliver services to the people by transforming its organisational set-up and the attitudes of officials. This process has not always been successful, remnants of the old administration are still prevalent and therefore dysfunctional. However, government administration has a number of advantages. All existing government institutions can be mobilised, and as Ahuja *et al.* have pointed out in their study, it is almost always better to use existing institutions than to create new ones. If a project has to introduce simultaneously a new technology, a new extension system, a new credit system, new production structures, etc., then failures are virtually assured.

A second advantage of government administration is that it can temporarily function as a buyer of the new technologies. Bulk orders are required to enable manufacturers to set up production lines and the government is the only institution that can buy the output as shown in various case studies. It should be stressed that this could be a temporary function as commercialisation would be feasible after some time.

A third advantage of government administration is that it can combine research functions with those of production and implementation. There are many research projects from which tested technologies can be selected and implemented. Only very large companies apart from the government are in a position to do this.

These advantages, in addition to the traditional role as a disburser of public funds, explain the relatively strong role of the government compared with that of NGOs and private manufacturers.

Although the advantages of employing government institutions for development projects have been recognised, it has also been pointed out that many shortcomings in the quality of government management are due to various constraints. In some countries this has led the government to farm out the management of projects to special authorities or even to commercial institutions, such as contractors and consultancy bureaus. In the case studies cited here this issue was of minor importance, but there are many examples in contemporary India where this kind of commercialisation of management has solved some of the above problems.

**Span of control**

The role of programmes and projects in introducing technical inputs in the process of rural transformation promoted by the Indian government has been considerable. Sometimes the programmes have been limited to a subsidy for manufacturers or users; in other cases an elaborate network of extension, credit, training for operation and maintenance has been set up. Here again the concept of 'span of control' can be used, which brings us to the third approach. In all the case studies, this span of control with which the management of the project has to function varies. To differentiate the span of control approach from the institutional approach, we could look at the functional limitations. In other words, not all the functions (or links in the chain of the project) are assumed to be covered by the project management. However, it should be able to collect information on—or monitor—these functions. If, for instance, credit or maintenance is not a function of a project, it is outside the span of control; but if they are still of strategic importance, they should be covered by monitoring. In the case studies cited here, this overall view was not present or at least not systematically developed.

However, this kind of approach is necessary and should be considered as a necessary condition for strategic management. Functions that produce bottlenecks or are not within the span of control of the project must be dealt with in one way or another. They can be absorbed into the project, for instance, by adding a credit scheme if funds are lacking. They can also be left to other institutions who can be encouraged to cover these functions. Instead of creating a new credit scheme, the existing bank credit schemes can be modified in order to accommodate the requirements of the project. But essentially, in both solutions, it is the monitoring system that has alerted the project management and stimulated it to action.

### Strategic management and project tactics

In the case studies, there is little strategic management in the sense in which it was defined in the introductory chapter. The explanation can be found in the constraints analysed here: lack of span of control, institutional limitations, and technological constraints. If these constraints are taken into account then the question arises: whether management could have done more within the limitations of the constraints? In other words, should we speak of good or bad management or of different management styles? We must conclude that on the basis of existing case studies these questions cannot be answered. More detailed case studies are required. What could be concluded is that the project design gave little scope for strategic management. The lesson is that projects should be designed to allow for strategic management. Strategies and tactics in the timing of the stages of projects are not very common. There are some exceptions such as the stove projects which were considered by NGOs to be an excellent instrument for mobilising women, and they were successful in doing so.

### Management styles

As mentioned in the introductory chapter, the management style of rural development projects has been strongly influenced by American concepts of project management. At the same time, the influence of the social-cultural context has been noted. Have these two factors been amalgamated into an 'Indian style' of management in the same way as Japan has developed its own management style? Who is a good manager in the setting as analysed in the various case studies?

In India itself there is a tendency to assign a negative value to the 'Indian' influence on management styles and managers. Deviations from 'Western' or 'American' management styles are considered undesirable. Nepotism and corruption are cited as examples of this influence that should be eliminated as soon as possible. On this specific point I would not disagree but, as pointed out in the introductory chapter, there is more to it than merely nepotism or corruption.

A good manager should be one who knows how to balance the more target oriented project approach and the network approach in which human relations both within and outside the project are defined by cultural codes. True, the word 'balance' is not easy to translate into indicators and judgement procedures. But the advantage of using it is the acceptance that there is an 'Indian style of management' in which the serving and maintaining of personal networks and the acceptance of the cultural codes that come with it is considered as an asset and not as a negative deviation from a 'Western' management style. The acceptance of particularistic approaches in the management of rural development projects is not without risks. If these approaches are dysfunctional in terms of the project objective, as Frank de Zwart maintains in his study on the transfer system, there can no longer be a question of balancing different value systems. However, meaningful balancing can be promoted by using mechanisms of public control. This public control requires proper monitoring and reporting with the public having access to this information, and an interest by the media in covering these projects. If the media are controlled by the local power elite not much can be expected from the control function and this will mean that the cultural code of investing in people can easily degenerate into straightforward nepotism and corruption.

# BIBLIOGRAPHY


**Abed, F.H.** (1983). Household Teaching of ORT in Rural Bangladesh, *Assignment Children*, Vol. 61–62.

**Abel-Smith, B.** (1985). Global Perspective on Health Service Financing, *Social Science and Medicine*, Vol. 21.

**Action** (1984). *Action Plan for Oral Rehydration in Gujarat State,* Government of Gujarat, Ahmedabad.

**Agarwal, Anil, Kimondo James, Gloria Moreno** and **Jon Tinker** (1981). *Water, Sanitation, Health for All? Prospects for the International Drinking Water Supply and Sanitation Decade, 1981–1990*, Earthscan, London.

**Agarwal, Bina** (1981). *Agricultural Modernisation and Third World Women: Pointers from Literature and an Empirical Analysis*, Working Paper Rural Employment Research Programme, ILO.

**Agarwal, Bina** (1986a). *Cold Hearths and Barren Slopes. The Woodfuel Crises in the Third World*, Allied Publishers, New Delhi.

**Agarwal, Bina** (1986b). Women, Poverty and Agricultural Growth in India, *The Journal of Peasant Studies*, Vol. 13, No. 4.

**Annual Report** (1985–86). *Annual Report 1985–1986*, Department of Non-Conventional Energy Sources, Ministry of Energy, Government of India, New Delhi.

**Bailey, G.V.J.** (1982). Tuberculosis Control in India. Current Problems and Possible Solutions, *Indian Journal of Tuberculosis*, Vol. 30.

**Banerji, D.** (1981). Social Aspects of the Tuberculosis Problem in India, in *Textbook of Tuberculosis*, Tuberculosis Association of India, New Delhi.

**Bapna, Ashok** (1985). Trends in India's Energy Consumption. A Search for Alternatives, in Rakesh Hooja & Yogeshwar Sharma eds., *Energy Alternatives for Rural Development*, Centre for Management Studies, Jaipur.

**Barua, B.N.M.** (1977). *Tuberculosis in India*, Tuberculosis Association of India, New Delhi.

**Basic** (1982). *Basic Information on Population and Family Program*, National Family Planning Coordination Board, Bureau of Population Data, Jakarta.

**Basu, R.N., Z. Jezek** and **N.A. Ward** (1979). *The Eradication of Smallpox from India*, New Delhi.

**Batliwala, Srilatha** (1983). Women and Cooking Energy, *Economic and Political Weekly*, Vol. 18, No. 52.

**Blankenberg, F.P.** (1983). *Implementation of Biogas Plants in Gujarat, India*, Working Paper No. 3, Technology and Development Group, Twente University of Technology, Enschede, The Netherlands.

**Blankenberg, F.P.** (1985). *Implementation of Rural Technology Projects in India: An Exploratory Research* (PhD Thesis), Twente University of Technology, Enschede, The Netherlands.

**Bose, Ashish** and **P.B. Desai** (1983). *Studies in Social Dynamics of Primary Health Care*, Institute for Economic Growth, New Delhi.

**Caincross, S., Ian Carruthers, Donald Curtis, Richard Feachem, David Bradley** and **George Baldwin** (1980). *Evaluation for Village Water Supply Planning*, John Wiley & Sons, Chichester.

**Cariappa, B.M.** (1981). Place of Tuberculosis Associations, in *Textbook of Tuberculosis*, Tuberculosis Association of India, New Delhi.

**Chauhan, Sumi** (1979). *A Village in a Million, an Energy Portrait of Akbarpur-Barota, India*, Earthscan, London.

**Chaudhury, Shekhar** (1985). *Strategic Management of Rural Technology Programmes in India. Renewable Energy, Afforestation, Rural Health and Rural Water Supply*, Draft Working Paper, Twente University of Technology, Enschede, The Netherlands.

**Contractor, A.** (1985). *Integrated Child Development Scheme: Going to Scale*, Paper presented at the Regional Workshop on Scaling-up for Child Survival, Indian Institute of Management, Ahmedabad.

**Contractor, A.A.** and **M.H. Shah** (1981). *Report on the Implementation of Expanded Programme on Immunisation, Gujarat, India*, Government of Gujarat, Ahmedabad.

**Contractor, A.A.** and **M.H. Shah** (1982a). *Coverage Evaluation Survey under EPI Programme, Kheda District, Gujarat, India*, Government of Gujarat, Ahmedabad.

**Contractor, A.A.** and **M.H. Shah** (1982b). *Background Paper on EPI, Training Programme for Mid-level Managers, Gujarat, India*, Government of Gujarat, Ahmedabad.

**Dasgupta, Biplab** (1971). *The Oil Industry in India, Some Economic Aspects*, Frank Cass & Company Ltd, London.

**D'Onofrio, Flores P.** and **S. Pfafflin** (1982). *Scientific-Technological Change and the Role of Women in Development*, United Nations Institute for Training and Research, Westview Press, Boulder.

**Dick, B.** (1985). Issues in Immunisation in Developing Countries, *EPC Publication*, No. 7.

**Djurfeldt, Goran** and **Staffan Lindberg** (1975). *Pills Against Poverty: A Study of the Introduction of Western Medicine in a Tamil Village*, Curzon Press, London.

**Eckholm, Erik, G. Foley, G. Barnard** and **L. Timberlake** (1984). *Fuelwood: The Energy Crisis That Won't Go Away*, Earthscan, London.

**Elias, Norbert** (1978). *The Civilizing Process. The Development of Manners; Changes in the Code of Conduct and Feeling in Early Modern Times (part II)*, Blackwell, Oxford.

**Foley, Gerald** and **Patricia Moss** (1985). *Improved Cooking Stoves in Developing Countries*, Technical Report No. 2, Earthscan, London.

**GAIC** (1982). *National Programme on Biogas Development, 6th Plan Period*, Gujarat Agro Industries Corporation, Ahmedabad.

**Gadgil, D.R.** (1973). *The Industrial Evolution in India in Recent Times 1860–1939*, Oxford, New Delhi (first published in 1924).

**GEDA** (no date). *Rural Energy Survey of 50 Villages in Gujarat, Summary*, Gujarat Energy Development Agency, Vadodara.

**Grant, J.P.** (1985). *The State of the World's Children*, UNICEF, New York.

**Gujarat** (1977). *Gujarat State Gazetteers, Kheda District*, Government of India, Ahmedabad.

**Gujarat** (1981). *Gujarat State Gazetteers, Banaskantha District*, Government of India, Ahmedabad.

**Gupta, Ranjit** (1981). The Poverty Trap: Lessons from Dharampur, in David C. Korten and Felipe B. Alfonso eds., *Bureaucracy and the Poor: Closing the Gap*, Asian Institute of Management, Manila.

**Health** (1980). *Health for All. An Alternative Strategy*, Indian Council of Social Science Research and Indian Council of Medical Research, New Delhi.

**Health Information** (1986). *Health Information of India, 1986*, Central Bureau of Health Intelligence, Directorate General of Health Services, New Delhi.

**Health Statistics** (1984). *Health Statistics of Gujarat*, Directorate of Health, Medical Services and Medical Education, Ahmedabad.

**Heginbotham, Stanley J.** (1975). *Cultures in Conflict: The Four Faces of Indian Bureaucracy*, Columbia University Press, New York.

**Henderson, R.H.** (1984). Immunisation Update: Recent Recommendations from the WHO Expanded Programme on Immunisation, *Contact*, No. 82.

**Hirschman, Albert** (1981). The Rise and Decline of Development Economics, *Development*, No. 3.

**Hooja, Bhupendra** (1985). Problems of Energy in Rural India, in Rakesh Hooja & Yogeshwar Sharma eds., *Energy Alternatives for Rural Development*, Jaipur.

**India's Environment** (1982). *The State of India's Environment 1982, A Citizens' Report*, Centre for Science and Environment, New Delhi.

**India's Environment** (1985). *The State of India's Environment 1984–1985. The Second Citizens' Report*, Centre for Science and Environment, New Delhi.

**Iyengar, I.** (1987). *Voluntary Agencies as a Link between the Rural Poor and the Credit Institutions*, Gujarat Institute of Area Planning, Ahmedabad.

**Jaggi, O.G.** (1979). *History of Science, Technology and Medicine in India. Volume Fourteen. Western Medicine in India: Public Health and its Administration*, Atma Ram & Sons, New Delhi.

**Jobert, B.** (1985). Population and Health Policy: The Case of Community Health Volunteers in India, *Social Science and Medicine*, Vol. 20, No. 1.

**Joseph, George, John Desrochers** and **Mariamma Kalathil** (1983). *Health Care in India*, Centre for Social Studies, Bangalore.

**Justice, Judithanne** (1983). The Invisible Worker: The Role of the Peon in Nepal's Health Service, *Social Science and Medicine*, Vol. 17, No. 14.

**Kelkar, Govind** (1981). *Impact of Green Revolution on Women's Work Participation and Sex Roles*, ILO Tripartite Asian Regional Seminar, Rural Development and Women, Mahabaleshwar.

**Kharabanda, V.P.** and **M.A. Quereshi** (1985). Giants Compared: Biogas Application Policy and Practice in China and India, *Development Forum*, Vol. 13, No. 10.

**Korten, D.C.** (1980). Community Organization and Rural Development: A Learning Process Approach, *Public Administration Review*, September–October.

**Kothari, Uttam, B. Mukarji** and **K.K. Paul** (1982). Health Services Utilization in Rural Areas, *Indian Journal of Pediatrics*.

**Mandl, P.E.** (1984). Going to Scale for Child Survival and Development, *Assignment Children*, Vol. 65–68.

**Mankodi, Kashyap and K.W. van der Veen** (1985). Treatment Failure in National TB Programme, *Economic and Political Weekly*, Vol. 20, No. 21.

**Maru, P.M.** (1985). The Community Health Volunteer Scheme in India: An Evaluation, *Social Science and Medicine*, Vol. 17, No. 19.

**Maternal** (1983). *Maternal and Childhood Information Gujarat (with Inserts for 1983–84 and 1984–85)*, Government of Gujarat, Ahmedabad.

**McCord, C.** (1976). *What's the Use of a Demonstration Project*?, Ford Foundation, Dacca.

**Mehta, K.N.** (1984). Status of Rural Water Supply Projects in Gujarat State, *Journal of Indian Water Works Association*, Vol. 16, No.1.

**Mies, Maria** (1981). *The Social Origins of the Sexual Division of Labour*, ISS, Den Haag.

**Morss, E.R.** and **J.K. Hatch** (1975). *Strategies for Small Farmers Development: An Empirical Study of Rural Development Projects*, Final Report prepared for the Agency for International Development, Washington.

**Madhi, S.S.** (1982). Biogas Programme in India, *Financing Agriculture*, Vol. 14, Nos. 2–3.

**Moulik, T.K.** (1982a). *Biogas Energy in India*, Indian Institute of Management, Ahmedabad.

**Moulik, T.K.** (1982b). Biogas System: Alternative Technology for Meeting Rural Energy Needs in India, *Financing Agriculture*, Vol. 14, Nos. 2–3.

**Moulik, T.K., P.R. Shukla** and **S.K. Sinha** (1985). *Optimal Selection from Centralized and Decentralized Energy Generating Technologies for Meeting Rural Energy Needs*, Indian Institute of Management, Ahmedabad.

**Moulik, T.K.** and **Shekhar Chaudhury** (1986). *Strategic Management of Rural Technology and Health Programmes in the Process of Rural Transformation: An Overview of the IDPAD Workshop*, Jaipur, 3–5 March.

**Moulik T.K.** and **M.A. Quereshi** (1984). *Strategic Management of Rural Technology Projects: The Case of Renewable Energy Technologies in Gujarat*, Indian Institute of Management, Ahmedabad.

**Mukerji, Mohan** (1982). *Story of a Chief Secretary During Emergency*, Ashok, New Delhi.

**Myrdal, Gunnar** (1970). *The Challenge of World Poverty: A World Anti-Poverty Program in Outline*, Vintage Books, New York.

**Nabrahman, D.** and **Shreekant Sambrani** (1983). Women's Drudgery in Firewood Collection, *Economic and Political Weekly*, Vol. 18, No. 1.

**Nagel, J.** (1985). Vaccines and Immunisation Schedules, *Tropical and Geographical Medicine*, Vol. 37, No. 3.

**Nagpaul, D.R.** (1981). Tuberculosis Control in Rural India, in *Textbook of Tuberculosis*, Tuberculosis Association of India, New Delhi.

**NIHFW (National Institute for Health and Family Welfare)** (1979). *Repeat Evaluation of Community Health Volunteers Scheme 1979. Survey Report. (Part 1)*, New Delhi.

**ORT** (1980). Oral Rehydration Therapy for Childhood Diarrhoea, *Population Reports*, Vol. 7, No. 6, John Hopkins University, Baltimore.

**Pai Panandiker, V.A.** (1982). *Personnel System for Development Administration*, Popular Prakashan, Bombay.

**Pamra, S.P.** (no date). *Lectures on Tuberculosis for General Practitioners*, The Tuberculosis Association of India, New Delhi.

**Pamra, S.P.** (1980). Problems of TB in Developing Countries. *Clinics in Chest Medicine*, Vol 1.

**Panchanadikar, K.C.** and **J. Panchanadikar** (1978). *Rural Modernisation in India. A Study in Developmental Structure*, Popular Prakashan, Bombay.

**Park, J.E.** and **K. Park** (1983). *Textbook of Preventive and Social Medicine*, (Ninth Edition), Banarsidas Bhanot, Jabalpur.

**Patel, Shri Tribhuvandas K.** (no date). *Kaira District. Mother and Infant Health Programme*, Tribhuvandas Foundation, Amul Dairy, Anand.

**Paul, Samuel** (1982). *Managing Development Programs, The Lessons of Success*, Westview Special Studies in Social, Political and Economic Development, Westview Press, Boulder.

**Pillai, S. Devadas** and **C. Baks** eds. (1979). *Winners and Losers. Styles of Development and Change in an Indian Region*, Popular Prakashan, Bombay.

**Pyle, D.** (1984). *Life After Project. A Multi-dimensional Analysis of Implementing Social Development Programmes at the Community Level*, John Snow Inc., Boston.

**Radakrishna, S.** (1983). National Tuberculosis Programme—Relative Merits of Enhancing the Operational Efficiency of Different Components of the Treatment Programme. *Indian Journal of Tuberculosis,* Vol. 30, No. 1.

**Raj, Maithreyi Krisna** (no date). *The Price of Gender Inequality is Technological Backwardness*, Research Unit on Women's Studies, SNDT Women's University, Bombay.

**Rajeswaran, K.** (1983). *Simple Energy Technology: Bio-gas Utilization in Tanzania*, Institute of Resource Assessment, University of Dar es Salaam.

**Ramachandran, P.G.** (1985). *Social Marketing. A Major Tool for Scaling-up Child Survival*, Paper presented at the Regional Workshop on Scaling-up for Child Survival Activities, Indian Institute of Management, Ahmedabad.

**Ranganiman, M.G., B.G. Prasad** and **V.K. Bhatnagar** (1968). Study of Medical Service Provided by PHC, *Indian Medical Care*, Vol. 6.

**Rao, K.N.** (1981). History of Tuberculosis, in *Textbook of Tuberculosis*, Tuberculosis Association of India, New Delhi.

**Rosaldo, Michelle Zimbalist** and **Louise Lamphere** (1974). *Women, Culture and Society*, Stanford University Press, Stanford.

**Rondinelli, D.A.** (1983). *Development Projects as Policy Experiments: An Adaptive Approach to Development Administration*, Methuen, London.

**Sarin, Madhu** (1984a). *How to Make and Use the Nada Chulha, Chulha Mistri's Manual*, Voluntary Health Association of India, New Delhi.

**Sarin, Madhu** (1984b). *Nada Chulha, a Handbook*, Voluntary Health Association of India, New Delhi.

**Sarin, Madhu** (1986). Improved Chulha Programme: Boon or Disaster, *Economic and Political Weekly*, Vol. 21, Nos. 38 & 39.

**Satia, J.K.** (1981). Developmental Tasks and Middle Management Roles in Rural Development, in David C. Korten & Felipe B. Alfonso eds., *Bureaucracy and the Poor: Closing the Gap*, Asian Institute of Management, Manila.

**Satia, J.K., D.V. Maylankar** and **I. Menon** (1985). *Scaling-up–Key Issues*, Background paper presented at the Regional Workshop on Scaling-up for Child Survival, Indian Institute of Management, Ahmedabad.

**Schenk-Sandbergen, Loes** (1984). Women's Work without Status: Outdoor Household Servants in Alleppey (Kerala), in Kenneth Ballhatchet & David Taylor eds., *Changing South Asia: Economy and Society*, SOAS, London, Asian Research Service, Hongkong.

**Schenk-Sandbergen, Loes** (1985). *Poor Rural Women and New Water and Fire Technologies in Gujarat (India)*, Report on an investigation, Department of South and Southeast Asia, University of Amsterdam/Indo-Dutch Programme on Alternatives in Development, The Hague.

**Schenk-Sandbergen, Loes** (1986). *People, Forests and Trees in Indian Society, Anthropological and Sociological Dimensions*, Annex 2, Report of the Mission of the Government of The Netherlands on the Identification of the Scope for Forestry Development Cooperation in India.

**Scott, James C.** (1972). *Comparative Political Corruption*, Prentice Hall, New Jersey.

**Sen, Gita** (1982). Women Workers and the Green Revolution, in Lourdes Beneria ed., *Women and Development, the Sexual Division of Labour in Rural Societies*, Praeger Publishers, New York.

**Shah, Ghanshyam** (1975). *Caste Associations and Political Process in Gujarat*, Popular Prakashan, Bombay.

**Shah, Ghanshyam** and **Biswaroop Das** (1988). *Voluntary Organizations and Development in Gujarat*, Centre for Social Studies, Surat.

**Sharma, Ursula** (1985). *Women, Work and Property in North-West India*, Tavistock, London.

**Simon, Herbert A.** (1976). *Administrative Behavior: A Study of Decision-Making Processes in Administraive Organization*, The Free Press, New York (first published in 1945).

**Skutsch, M.M.** (1986). *Strategic Management of Stove Programmes: A Rationale Reconstruction Approach*, Technology and Development Group Working Paper, No. 25, Twente University of Technology, Enschede, The Netherlands.

**Skutsch, M.M.** (1987). *The Not-So-Social Forestry in Gujarat in India*, Technology and Development Group Working Paper, No. 29, Twente University of Technology, Enschede, The Netherlands.

**Sokhey J., J. Bhargava** and **R.N. Basu** (1984). *The Immunisation Programme in India, a Handbook for Medical Officers*, Ministry of Health and Family Welfare, New Delhi..

**Srinivas, M.N.** (1955). The Social System of a Mysore Village, in McKim Marriott, *Village India. Studies in the Little Community*, The University of Chicago Press, Chicago.

**Statistical Outline** (1987). *Statistical Outline of Gujarat, 1986*, Bureau of Economics and Statistics, Gandhinagar.

**Streefkerk, H.** and **P. Streefland** (1980). *Brackish Water and Poor People, Report on a Short Mission to Banaskantha District, Gujarat State, India*, Royal Tropical Institute and University of Amsterdam, Amsterdam.

**Streefland, P.** and **C.M. Varkevisser** (1985). *Variations in Primary Health Care*, Paper presented at the International Symposium on Effectiveness of Rural Development Cooperation, Royal Tropical Institute, Amsterdam.

**Subramanian, Ashok** (1983). *Managing Voluntary Health Organisations*, Paper prepared for a Workshop of Voluntary Agencies, Indian Institute of Management, Ahmedabad.

**Tanothai, A.** (1985). *Population and Community Development Association, Thailand*, Paper presented at the Regional Workshop on Scaling-up for Child Survival, Indian Institute of Management, Ahmedabad.

*Textbook of Tuberculosis* (1981). The Tuberculosis Association of India, New Delhi.

**United Nations Conference** (1981). *United Nations Conference on New and Renewable Sources of Energy, National Paper India*, Nairobi.

**Veen, K.W. van der** (1981). Socio-cultural Aspects of Medical Care in Valsad District, Gujarat State, in Giri Raj Gupta ed., *The Social and Cultural Context of Medicine in India. Main Currents in Indian Sociology, Vol. IV*, Vikas, New Delhi.

**Veen, K.W. Van der** (1982). *Socio-cultural Factors in TB Care. A Case Study in Valsad District, Gujarat, India*, Working Paper of the Department of South and Southeast Asian Studies, No. 8, University of Amsterdam.

**Viegas, Philip** and **Geeta Menon** (1985). The Social Costs of Deforestation, *Social Action*, Vol. 35.

**Wade, R.** (1982). The System of Administrative and Political Corruption: Canal Irrigation in South India, *The Journal of Development Studies*, Vol. 18, No. 3.

**Wade, R.** (1985). The Market for Public Office: Why the Indian State is not Better at Development, *World Development*, Vol. 13, No. 4.

**Wertheim, W.F.** (1963). Sociological Aspects of Corruption in Southeast Asia, *Sociologica Neerlandica*, Vol. 1, No. 2.

**World Health Organisation** (1981). *Objectives and Targets, Training Course for National CDD Managers, CDD Programme*, WHO, Geneva.

**World Health Organisation** (1983). *Principles of Diarrhoeal Diseases Treatment, Training Course for Peripheral Health Workers, CDD Programme*, WHO, Geneva.

**Zwart, F. de** (1985). *The Function of Personnel Transfers in Indian Bureaucracy: A Case Study in a Public Health Department* (MA Thesis), Anthropological-Sociological Centre, University of Amsterdam.

# CONTRIBUTORS

**Kanta Ahuja** is Professor, University of Rajasthan, Jaipur.

**Floris Blankenberg** is at NOVIB, The Hague.

**Deepak Gianchandani** is at the Institute of Development Studies, Jaipur.

**Enno Hommes** is Professor, Twente University of Technology, Enschede, The Netherlands.

**T.K. Moulik** is Professor, Indian Institute of Management, Ahmedabad.

**J.K. Satia** is Professor, Indian Institute of Management, Ahmedabad.

**M.S. Rathore** is at the Institute of Development Studies, Jaipur.

**Loes Schenk-Sandbergen** is at the Anthropological-Sociological Centre, University of Amsterdam, Amsterdam.

**Margaret Skutsch** is at the Twente University of Technology, Enschede, The Netherlands.

**Hein Streefkerk** is at the Anthropological-Sociological Centre, University of Amsterdam, Amsterdam.

**Pieter H. Streefland** is Professor, Royal Tropical Institute, Amsterdam and University of Amsterdam, Amsterdam.

**Klaas W. van der Veen** is at the Anthropological-Sociological Centre, University of Amsterdam, Amsterdam.

**Frank de Zwart** is at the Department of Public Administration, University of Leiden, Leiden, The Netherlands.

# INDEX